SpringerBriefs in Electrical and Computer Engineering

More information about this series at http://www.springer.com/series/10059

Ivica Kuzmanić · Igor Vujović

Reliability and Availability of Quality Control Based on Wavelet Computer Vision

 Springer

Ivica Kuzmanić
Faculty of Maritime Studies
University of Split
Split
Croatia

Igor Vujović
Faculty of Maritime Studies
University of Split
Split
Croatia

ISSN 2191-8112 ISSN 2191-8120 (electronic)
SpringerBriefs in Electrical and Computer Engineering
ISBN 978-3-319-13316-4 ISBN 978-3-319-13317-1 (eBook)
DOI 10.1007/978-3-319-13317-1

Library of Congress Control Number: 2014956370

Springer Cham Heidelberg New York Dordrecht London

Printed on acid-free paper

Springer International Publishing AG Switzerland is part of Springer Science+Business Media
(www.springer.com)

Contents

Abstract

This work offers a framework for the assessment of reliability and availability of visual quality control systems explained on a wavelet computer vision system. The work presents experimental results pertaining to the sensitivity of visual quality control to noise, as an example of dependences. The reliability analysis of the proposed visual quality control system is presented in detail. The influencing parameters are analyzed and included in the reliability model. These parameters are divided into the software and the hardware group, with a condition representing a combination of software and hardware, and a condition representing the combination of hardware and environmental conditions. In the end, possible alternative approaches are suggested. System availability and reliability models were presented and solutions calculated.

Abstract

Chapter 1
Introduction

Abstract This chapter presents the contents of the book. A literature overview of the topic is presented as an introduction to the book's topic.

Keywords Quality control · Image quality assessment · MTBF · Reliability analysis

Due to the increasing demand for quality control systems more and more quality control systems are being proposed and published every year. The system reliability analysis could not keep up with the increasing demand. Due to the scarcity of such research, it would be interesting to know whether a particular quality control system is acceptably reliable and available. In this book we present such an analysis of our visual quality control system based on wavelets, in overall introduced in [1]. This book presents a framework, with the possibility of application of a similar approach to any quality control system.

The importance of reliability analysis is outlined in many references, i.e. in [2–4]. Reliability of constituting components was evaluated in several references, i.e. in [5–10]. The overall system reliability and availability consisting of three components was elaborated in [11]. The importance of reliability analysis for the maintenance and optimization of spare part inventories is discussed in [12]. This research begun with the use of wavelets as diagnostic tools in the turning process [13], which resulted in the detection of damage to the surface of inserts. The logical progress was to differentiate between damaged and good inserts. Wavelets are well-known mathematical tools and explanations can be found in many references, i.e. in [14–16].

This book is organized as follows. Some mathematical introductions and definitions are presented in Chap. 2. Computer vision quality control system is described in Chap. 3. The experimental results of noise influence are presented in the same chapter. Parameters influencing system performance are discussed in the Chap. 4. Reliability analysis is elaborated in Chap. 5. The final chapter presents conclusions and guidelines for further research.

© The Author(s) 2015

I. Kuzmanić and I. Vujović, *Reliability and Availability of Quality Control Based on Wavelet Computer Vision*, SpringerBriefs in Electrical and Computer Engineering, DOI 10.1007/978-3-319-13317-1_1

1.1 Literature Overview

Reliability analysis is the subject of many references from different fields of application. Software is notorious for being a part of the computer system which often fails. Even in early researches [17], models for reliability software analysis were investigated. Computer systems were analyzed for reliability in [18]. This analysis incorporates Markov model. The availability and reliability analysis of repairable systems was explained in [19]. Markov chains and time varying failure rates were combined in [19].

Communication protocols and network reliability in industrial applications were presented in [20]. The distribution of real-time over the network is important for image processing and camera transmissions should be monitored and analyzed. The network consists of the Ethernet, fire-wall and distributed data system which ensure connection to the imaging control station. Communication is established through the modem device. The real-time and delay time properties were considered. Robot vision in quality control process was considered in [21]. The influence of illumination on the accuracy of quality control constitutes a part of the chapter. An example considered was electromotor stator assembly.

Real applications of computer vision, which include variations in operating conditions, result in poor reliability [22]. Consequently, real world applications require frequent re-setup or re-initialization. Therefore, it is imperative to use some sort of a self-configurating, self-repairing, error detecting and recovery algorithm or similar. A software model was presented which uses techniques for regulation of internal parameters, error detection and recovery, self-configuration and self-repair for vision systems.

Like in our book, [23] deals with quality assessment of image quality. The proposed algorithm is available in [24] and metrics in [25].

Since we use communication in our research, it is interesting to mention that an availability and reliability analysis of a communication system was presented in [26]. Our research and [26] differ in methodology and approach. Furthermore, it is irrelevant for our research whether communication is outdoor or indoor, because it deals with the results of system performance. Communication in our research is merely a part of the system.

The use of computer vision in real-time is an intuitive choice in many industrial applications. Due to the fact that computer vision systems produce a large matrix of measured, often noisy data at regular sampling intervals, vision algorithms are stochastic and unpredictable [27]. This makes it difficult to understand the processes described by the data matrix and makes generalizations impossible. Examples of industrial applications can be found in [28–30]. However, these references did not include reliability and/or availability analysis.

The reliability and availability analysis is taken more seriously in an increasing number of applications each day. For example, mobile robots were considered in [31]. It is concluded that MTBF (Mean Time Between Failures) is 8 h and availability less than 0.5 (50 %). As green technology is a trend, the reliability of such systems is also considered. An example of reliability of wind turbines is in [32]. Reference [32] covers practical methods for predicting large wind turbine reliability using grouped survey data, as a contribution to green energy research.

Chapter 2
Mathematical Background

Abstract This chapter provides a mathematical background for understanding of the mathematical tools used in the book: basic definitions of main terms, Markov model basics, and references for wavelets background.

Keywords Markov model · State transition · MTBF · MTTR · 2D DWT

In this chapter, we provide the basic definitions and the mathematical background necessary for understanding the context.

2.1 Definitions of Reliability and Availability

Availability is defined by [11]:

$$A_i = \frac{MTBF}{MTBF + MTTR} \tag{2.1}$$

where $MTTR$ is Mean Time To Repair and Mean Time Before Failure ($MTBF$). Availability is often expressed as:

$$A_i = \frac{\mu}{\mu + \lambda} \tag{2.2}$$

where λ is the intensity of failures and μ the intensity of repairs. Intensity of failures can be determined by [11, 33]:

$$\lambda = \frac{1}{MTBF} \tag{2.3}$$

Intensity of repairs is defined with [11]:

I. Kuzmanić and I. Vujović, *Reliability and Availability of Quality Control Based on Wavelet Computer Vision*, SpringerBriefs in Electrical and Computer Engineering, DOI 10.1007/978-3-319-13317-1_2

$$\mu = \frac{1}{MTTR} \qquad (2.4)$$

2.2 Markov Models

The reliability and availability of a system can be established after an in-depth evaluation of the system. Some aspects meriting attention are:

- configuration of elements in the system,
- modes of operation of the system,
- component failure processes,
- conditions indicating that the system failed, and
- reparability of the system.

If the system is in one of the finite number of states in the observer time instant and if components fail in stochastic manner, the reliability and availability of the system can be established with the Markov theory.

Markov's models are functions or two variables:

- states of the system, $X(t)$, and
- observation time, t.

Both can be discrete or continuous in time. Based on the type of variables, Markov models can have four different forms:

- both variables are of discrete type,
- both variables are of continuous type,
- $X(t)$ is continuous and t discrete, and
- $X(t)$ is discrete and t continuous.

Markov models are called Markov chains if t is discrete. Markov models are called Markov processes if t is continuous. They depend on a set of probabilities, p_{ij}, indicating the transition of the system from state i to state j. A special case in the Markov process, interesting from the reliability and availability point of view, is the Poisson's process, which is, in fact, a model with discrete system's states and continuous time.

Equations, such as those presented below can be obtained from the table of transitions:

$$P_0(t + \Delta t) = (1 - \lambda_{01}\Delta t)P_0(t) \qquad (2.5)$$

$$P_1(t + \Delta t) = \lambda_{01}\Delta t P_0(t)(1 - \lambda_{12}\Delta t)P_1(t) \qquad (2.6)$$

$$P_2(t + \Delta t) = \lambda_{12} \Delta t P_1(t) + P_2(t) \tag{2.7}$$

If we put limit for $\Delta t \to 0$, we get differential equations, i.e.:

$$\frac{dP_0(t)}{dt} + \lambda P_0(t) = 0 \tag{2.8}$$

$$\frac{dP_1(t)}{dt} + \lambda_{12} P_1(t) - \lambda_{01} P_0(t) = 0 \tag{2.9}$$

$$\frac{dP_2(t)}{dt} - \lambda_{12} P_1(t) = 0 \tag{2.10}$$

By applying Laplace transform, the set of differential equations is transformed into a set of algebraic equations in the s-domain:

$$(s + \lambda_{01}) L[P_0(t)] = P_0(0) \tag{2.11}$$

$$-\lambda_{01} L[P_0(t)] + (s + \lambda_{12}) L[P_1(t)] = P_1(0) \tag{2.12}$$

$$-\lambda_{12} L[P_1(t)] + s L[P_2(t)] = P_2(0) \tag{2.13}$$

By solving equations and taking the inverse Laplace transform, we reach the solution for the desired probabilities:

$$P_0(t) = e^{-\lambda_{01} t} \tag{2.14}$$

$$P_1(t) = \frac{\lambda_{01}}{\lambda_{12} - \lambda_{01}} \left(e^{-\lambda_{01} t} - e^{-\lambda_{12} t} \right) \tag{2.15}$$

$$P_2(t) = 1 - \frac{\lambda_{12}}{\lambda_{12} - \lambda_{01}} e^{-\lambda_{01} t} + \frac{\lambda_{01}}{\lambda_{12} - \lambda_{01}} e^{-\lambda_{12} t} \tag{2.16}$$

In simplified cases, we can define three states for a two-component system:

- state S_0, in which both components are operating correctly,
- state S_1, in which one of the components is not operating correctly, and
- state S_2, when both components are malfunctioning.

The relationship between these three states can be graphically represented as in Fig. 2.1. Figure 2.1 shows that the intensity of failures leads to the next state and the intensity of repairs to the previous state and values are probabilities for the considered time period.

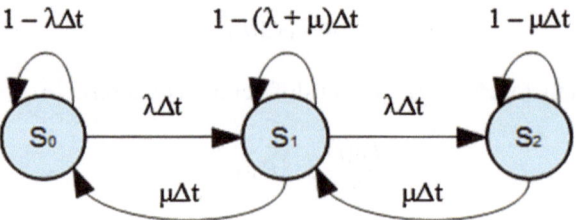

Fig. 2.1 Graphical representation of states' transitions [11]

2.3 Wavelet Role in Image Processing

Wavelets derive their strength from the Heisenberg uncertainty principle [34–36], but there are misunderstandings. Wavelets are, basically, used for non-stationary signal processing in the time-frequency domain [37–40]. The change of frequency and time resolution is possible due to the trade-off within the constraints of the Heisenberg principle. However, there are several ambiguities when dealing with images [41–43]. First, which value to assign to frequency and which to time. Furthermore, since the images are multidimensional we have to know the position of a pixel and one or more color values. When dealing with video, the time when the frame is taken is also of importance [44]. First, the wavelet transform (WT) is defined by mathematical expressions. Since the meaning of a particular parameter in the physical world is irrelevant, any physical parameter can be scaled or translated through WT. The only thing that matters is which physical parameter can be called frequency or time in a particular application.

Two-Dimensional Discrete WT (2D-DWT) takes care of pixel position and color values or intensity (in case of gray image), where the first parameter is location and the second color [34, 39, 41]. So, we localize frequency of color occurrence in an image. This property can be used in different applications, from histogram, edge detection or shadow detection to advanced computer vision algorithms.

When dealing with image sequences, videos, we must bear discrete time in mind. Every time instant corresponds to a single frame or the entire matrix with 2 coordinates of the pixel and values of the pixel (colors or intensity). Such data are commonly saved in multidimensional structures.

Wavelet can be used in preprocessing of images of visual quality control, such as denoising of the input data. Figure 2.2b, c shows an example of denoising of the source image Fig.2.2a. Figure 2.2b shows noised image and Fig. 2.2c denoised.

Another application is in the core of the algorithm, such as in feature extraction. Figure 2.2d, e shows an example of feature extraction—edge detection with Canny edge detector and gradient wavelet method. Figure 2.2d shows results of Canny edge detection for the sourced image.

The third possible application of wavelets is in post-processing, including archiving or image compression. Figure 2.2f, g, h shows compressed image at different levels in multiresolution compression.

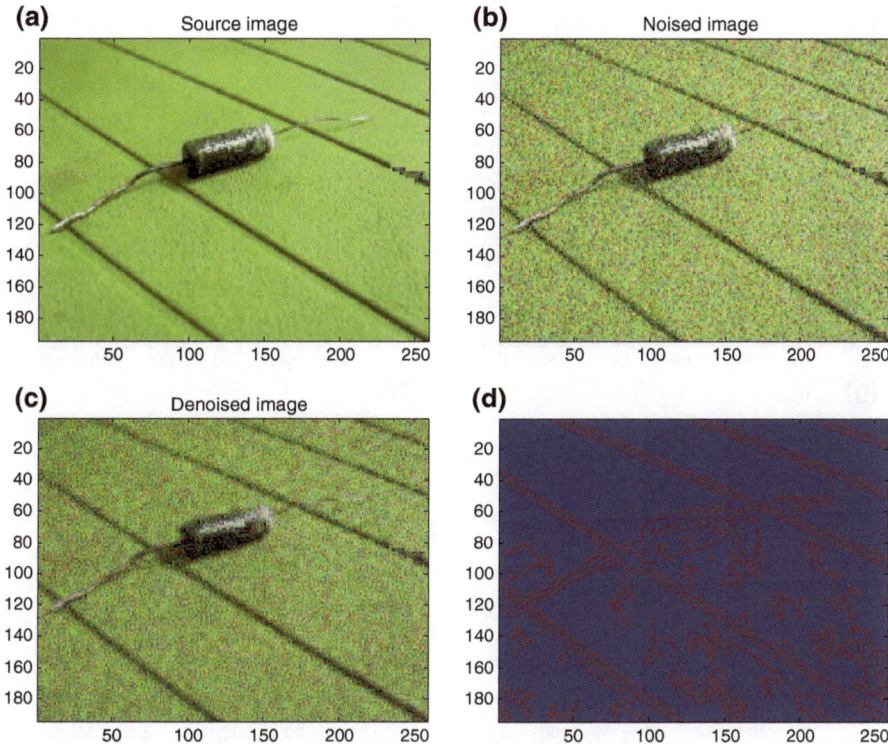

Fig. 2.2 Usage of wavelets: **a** source image, **b** noised source image, **c** denoised image, **d** Canny edge detection, **e** wavelet edge detection, **f** 1st level of the wavelet compression, **g** 2nd level of the wavelet compression, **h** 3rd level of the wavelet compression

Several novel transforms were presented in a past few decades, which poses enhanced characteristics for image processing purposes. Some are better for curves or edges and some for sharpness, compression, noise removal, etc. Some of these transforms are:

- curvelets [45],
- wedgelets [46],
- shapelets [47],
- bandelets [48],
- edgelets [49], etc.

Furthermore, WT is improved by introducing complex numbers [50, 51]. One should always keep in mind what are the benefits of advanced transforms usage, or the ration of gains and losses. For example, some transform can exhibit better characteristics, but the execution can take more time than it can be spared in some application. Therefore, such transform is useless in practice.

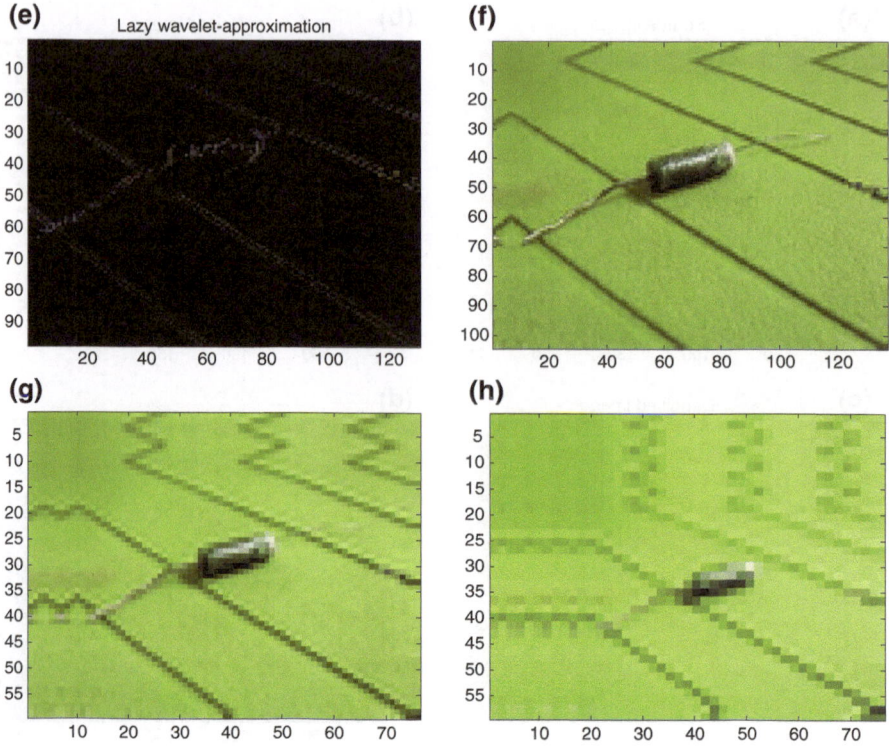

Fig. 2.2 (continued)

For this research, we used classic discrete WT in two dimensions (2D-DWT) with filter implementation, because it was not necessary to use more complex transforms for noise removal.

Chapter 3
A Model of Quality Control Computer Vision System

Abstract In this chapter, a visual quality control system is described. The system is based on a stationary camera. The experiment tested the influence of noise to the performance of the system. Three frequently used noises were selected for the test: Gaussian, salt and pepper, and speckle.

Keywords Visual quality control · Noise · Action phase · Observe phase

The analyzed quality control system is illustrated in Fig. 3.1. The system's input is a stationary camera for the analyzed example. The camera can be placed in a robotic arm, but that involves some technical issues which can make things more complex and are not necessary for the basic idea of the article. When the system is in "observe phase" (Fig. 3.1a), it observes the product passing underneath it.

A fixed camera forwards the video input to the computer, which analyses the quality of the product. If the software detects an unsatisfactory part of the product, the system is triggered into "action phase". In the "action phase" (Fig. 3.1b), the communication interface sends the action order to the actuator, which throws away the unsatisfactory piece. If the piece is of satisfactory quality, then the "action phase" is skipped.

The control software takes into account the fact that the pieces of the product are moving. It is important for the reaction time to be short enough to allow the removal of the undesired piece.

It is important for the system to operate under noisy conditions, because the conditions cannot always be expected to be ideal. The ability to distinguish satisfactory from unsatisfactory pieces is tested by adding noise. The system should be able to distinguish between satisfactory and unsatisfactory products under noisy conditions. Therefore, the first test was conducted in a noise-free environment and the following three with artificially produced noise. Figure 3.2 shows a sample of tests in different noisy conditions.

Three frequently used noises were selected for the test: Gaussian white of zero mean and 0.01 variance, salt and pepper with noise density 0.05 and speckle. In speckle noise experiment, multiplicative noise is added, which follows the equation:

I. Kuzmanić and I. Vujović, *Reliability and Availability of Quality Control Based on Wavelet Computer Vision*, SpringerBriefs in Electrical and Computer Engineering, DOI 10.1007/978-3-319-13317-1_3

Fig. 3.1 Analyzed quality control system based on computer vision: **a** phase of observation, **b** phase of action

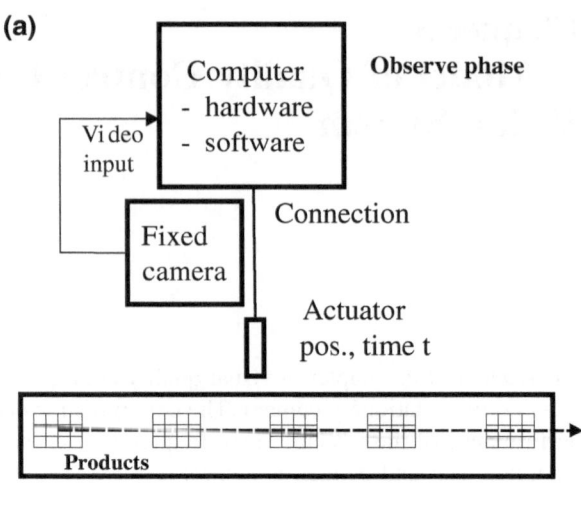

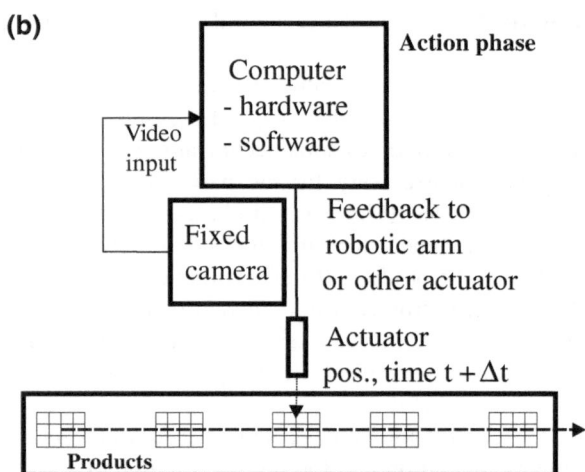

$$I_{noise} = I + n \cdot I = I \cdot (1 + n) \qquad (3.1)$$

where I_{noise} is the image with noise, I clear image, n is uniformly distributed random noise. Noise has the mean 0 and variance $v = 0.04$.

Table 3.1 shows results of the system evaluation. The results were obtained from the experiment designed in the research presented in this book. The wavelets were observed to increase tolerance to error. In the original image, it is only 1 %, while the used wavelets are best suited for speckle noise. This means that the image of the product can be changed by as many as 25 % (in case of speckle noise) without influencing the correctness of product classification (for removal or not).

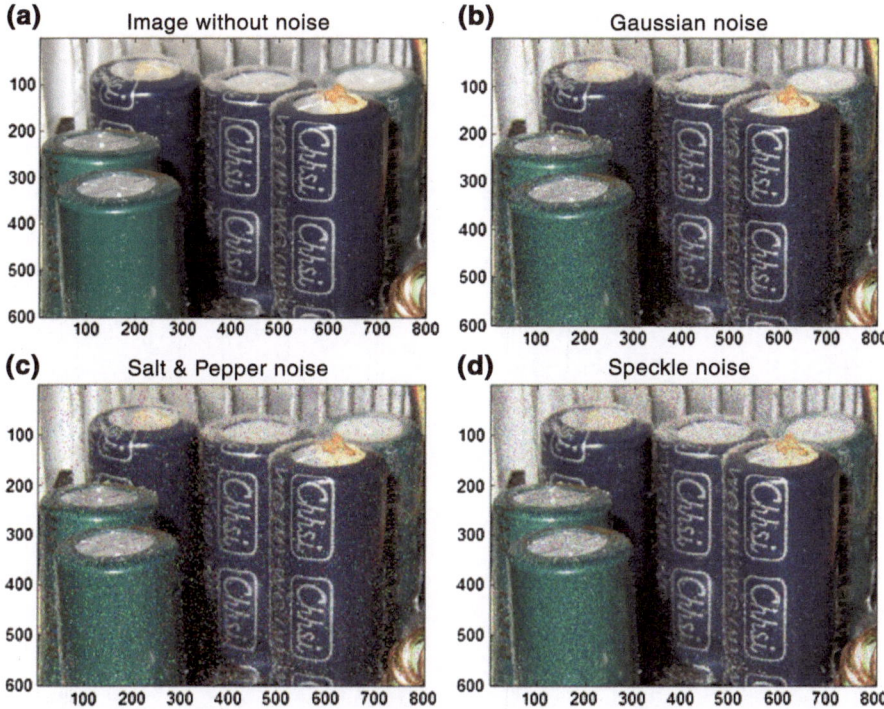

Fig. 3.2 **a** Image of the product, **b** product image with added Gaussian noise, **c** image with added salt and pepper noise, **d** image with added speckle noise

Table 3.1 Performance of the visual quality control system—experimental results

Clear/ noisy	Normalized average ROI energy for satisfactory product	Normalized average ROI energy for unsatisfactory product	Detection or recognition margin	Tolerance to error (%)
Clear image	0.5872	0.5813	0.0059	1.005
Gaussian noise	0.6115	0.5780	0.0335	5.478
Salt and pepper noise	0.5517	0.5374	0.0143	2.592
Speckle noise	0.6743	0.5046	0.1697	25.167

Noise is the most interesting influencing parameter, which has a direct impact on product classification. In the next chapter, we will consider other influencing parameters.

Chapter 4
Parameters Influencing Reliability and Availability of the System

Abstract Parameters influencing reliability and availability are considered in this chapter. Parameter's influence is analyzed by binarization of the final result: 0 if the result is unsatisfactory, and 1 if the result is satisfactory. Hardware and software parameters are considered.

Keywords Hardware parameters · Environmental parameters · Software parameters · Binarization

Influencing parameters can be grouped in two sets: hardware and environment parameters and software parameters.

The first hardware/environment parameter influencing a system's performance, is camera noise and camera quality and operation. It is peripheral hardware, and its state is designated with S_1. If something is wrong, $S_1 = 0$, and if everything is right, $S_1 = 1$.

The second hardware/environment parameter deals with scene/indoor environment characteristics. It is not the same if the factory hall is full of smoke or dust and if the hall is sterile. The state, S_2, is marked as 0 if scene is too dark, too dusty, or similar. In case there are no shadows and everything is fine, $S_2 = 1$.

The third parameter deals with communication equipment, such as wires and optical fibers (in case of non-wireless communications), fieldbus, profibus, ethernet or other. If the connection/link is functioning correctly, $S_3 = 1$ and otherwise $S_3 = 0$. This state partially involves software, because hardware communication equipment can function correctly while software or protocols may simultaneously cause communication errors.

The next parameter deals with computer hardware failure (which includes memory, HDD and/or processor failure) [11]. This parameter will be marked as S_8. If the hardware is functioning properly, then $S_8 = 1$, otherwise $S_8 = 0$.

Available processing time, S_7, is a mixed parameter (hardware and software). It depends on hardware execution time and software complexity. If the software is less demanding, even slower processor can perform the designed functions rapidly and on time. Therefore, S_7 can be divided into two components: S_{7A} and S_{7B}. $S_{7A} = 1$ for a sufficiently fast algorithm, and 0 otherwise. $S_{7B} = 1$ if an adaptation is

© The Author(s) 2015

I. Kuzmanić and I. Vujović, *Reliability and Availability of Quality Control Based on Wavelet Computer Vision*, SpringerBriefs in Electrical and Computer Engineering, DOI 10.1007/978-3-319-13317-1_4

made which allows the execution of the algorithm on a particular hardware in available time or if hardware is fast enough for the desired algorithm. It can be written:

$$S_7 = S_{7A} \,|S_{7B} \tag{4.1}$$

Software parameter (S_4), which influences the system performance, is the choice of a wavelet or wavelet tree. If satisfactory wavelet is selected and a good part of the wavelet tree, $S_4 = 1$, otherwise it is set to 0.

Everything can be fine (hardware and software), but the algorithm can still give unsatisfactory results due to two possible reasons: threshold selection (S_5) or reference model quality (S_6). If the threshold selection process produces a satisfactory result, than $S_5 = 1$, otherwise S_5 is set to 0. If the reference model is of sufficient quality, then $S_6 = 1$, otherwise it is set to 0.

Formally, there are two states of the quality control system, which should be introduced as well: the state without fault (S_0) and the state of failure (S_F).

Chapter 5
Analyzing and Modeling Reliability and Availability of the Quality Control System

Abstract First, it is considered whether this system can be considered as the Poisson's process. Both the reliability and the availability models are developed and explained parameter by parameter in detail (for the system introduced in the previous chapter). Solutions of the state differential equations system are found for the reliability and partial solution for the availability. Extended procedure of how to obtain the obtained solutions is presented in the appendix at the end of the book.

Keywords Reliability model · Availability model · Differential equations · Probability of transition

Reliability is actually the probability that our system or subsystem will successfully perform the desired task in a specific time period within operational constrains. Usually, researchers use the Poisson process to describe the availability and reliability of the system. The Poisson process is an interesting instance of the Markov process, which has discrete states, $X(t)$, and continuous time, t. Since most problems are solved by the Poisson assumption, we will first consider whether our process is a Poisson process.

Definition 5.1 The process is Poisson's if four assumptions are met:

1. the probability of transition from state with i events to state with $i + 1$ events over time interval Δt is equal to the product of constant λ and time interval Δt, which is $\lambda \Delta t$. Parameter λ designates events in a unit of time. From the viewpoint of reliability, parameter λ is the intensity of failures;
2. all events are independent;
3. events are irreversible, which means that failure rate increases over time in the case of irreparable systems. In this event there is no maintenance after failure and consequently, there is no chance of return to the previous state.
4. The probability of occurrence of two or more events in interval Δt is negligible.

Observation 5.1 It is a matter of dispute whether conditions from Definition 5.1. have been met. For example, if hardware fails, software cannot perform. Therefore, although the software may be functioning properly, the result may be unsatisfactory

© The Author(s) 2015

I. Kuzmanić and I. Vujović, *Reliability and Availability of Quality Control Based on Wavelet Computer Vision*, SpringerBriefs in Electrical and Computer Engineering, DOI 10.1007/978-3-319-13317-1_5

due to hardware failure. So, software states may be said to depend on hardware states. Therefore, it can be concluded that our process is not Poisson's. However, strictly speaking, the software is functioning properly. For example, edge detection can perform correctly based on input. It is not the fault of the edge detector if the input image has bad or even unacceptable noise levels or degradation of data. Edge detector performed as well as can be expected with deteriorated data. So, the edge detector may be said to have properly functioned and the data acquisition hardware poorly. It is unclear whether this is Poisson's process or not.

5.1 Reliability Analysis

In order to develop a reliability model, we need to include considerations from Chap. 3. Furthermore, we need also consider interdependences between possible failures. For example, a bad reference model will influence the reliability of the threshold or wavelet selection process. Link failure will result in the malfunction of the entire system, etc.

Figure 5.1 illustrates the Markov model of reliability analysis of the system. The basic state is state S_0, which signifies that the system is functioning properly. If the system does not function or functions with unsatisfactory results, it is in a faulty state, S_F. Figure 5.1 shows possible paths from S_0 to S_F. It also shows interactions between states. S_F can be obtained if any of the S_i is in a faulty state (zero value). Some states interfere with other states (dashed line) and theirs condition influences

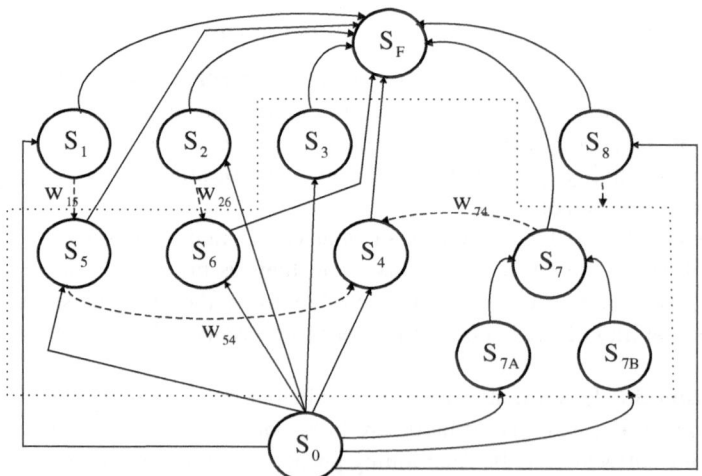

Fig. 5.1 Markov model representation for reliability

the condition of the linked state with some weight (w_{ij}). Other states influence the final state with $w_{iF} = 1$ (100 %).

Weight w_{15} shows how failure in S_1 influences the S_5 state. It means that unsatisfactory functioning of the camera influences the threshold selection. In the reliability analysis, we identified several interdependences: w_{15}, w_{26}, w_{54}, w_{74} and S_8 influences S_3, S_4, S_5, S_6, and S_7 with w_{83}, w_{84}, w_{85}, w_{86} and w_{87}. If computer hardware fails, the results of the proper functioning of other system parts are irrelevant. For example, if camera operates correctly and a computer processor malfunctions, the entire system does not work. Therefore, if $S_8 = 0$, then associated weights propagate the error through the system with 100 % and therefore $w_{83} = w_{84} = w_{85} = w_{86} = w_{87} = 1$.

However, the influence of the camera to threshold selection is harder to quantify, but it is obvious that camera quality and operation influences threshold selection. Therefore, w_{15} is difficult to estimate. Since this consideration is theoretical, weight values are irrelevant. In practical calculations, the actual weight values need to be carefully determined.

The next step is to analyze the system and failures state by state. For example, when one component fails, it can cause another component to experience failure as well. However, since this system is not redundant, if any component fails, the entire system changes state to fault state. Therefore, the state transition relation is:

$$S_0 \rightarrow S_i \rightarrow S_F \qquad (5.1)$$

When the system changes state to a fault state, it cannot repair itself and reverts to initial state. Therefore, from any S_i, the system may only change state to fault state, S_F, in future time instances. When the system is repaired, it changes state to initial state, S_0. The probability of such an event is marked with r_{F0} in Table 5.1. According to Table 5.1, $p_{S0F} = 0$. The probability that a faulty system will repair itself is equal to zero ($r_{FF} = 1$).

Table 5.1 Reliability table—relationship between initial state and final state (transitions)

$P_S(t)$—initial state	$P_S(t+\Delta t)$—final state									
	P_{S0}	P_{S1}	P_{S2}	P_{S3}	P_{S4}	P_{S5}	P_{S6}	P_{S7}	P_{S8}	P_{SF}
P_{S0}	r_{00}	r_{01}	r_{02}	r_{03}	r_{04}	r_{05}	r_{06}	r_{07}	r_{08}	0
P_{S1}	0	r_{11}	0	0	0	r_{15}	0	0	0	r_{1F}
P_{S2}	0	0	r_{22}	0	0	0	r_{26}	0	0	r_{2F}
P_{S3}	0	0	0	r_{33}	0	0	0	0	0	r_{3F}
P_{S4}	0	0	0	0	r_{44}	0	0	0	0	r_{4F}
P_{S5}	0	0	0	0	r_{54}	r_{55}	0	0	0	r_{5F}
P_{S6}	0	0	0	0	0	0	r_{66}	0	0	r_{6F}
P_{S7}	0	0	0	0	r_{74}	0	0	r_{77}	0	r_{7F}
P_{S8}	0	0	0	r_{83}	r_{84}	r_{85}	r_{86}	r_{87}	r_{88}	r_{8F}
P_{SF}	r_{F0}	0	0	0	0	0	0	0	0	r_{FF}

Table 5.1 shows the transient matrix for all changes of state. If a change is impossible, the corresponding table element is equal to zero. It can be seen that transitions from initial state S_0 to any fault state S_i are possible, with probability r_{0i}. When a system changes its state from S_0 to S_i, it may remain in S_i for some time before transitioning to fault state S_F. Table 5.1 shows reliability analysis transitions until the first fault and the expected change to the more permanent S_F state. After repairs, the system changes state to S_0 and everything can be repeated.

If the system was already in state S_1, the damage may be increased if it transitions to state S_5 before turning to S_F.

It should be kept in mind that r_{ij} are probabilities, not state values. In the case of state values, we assumed binary functions—0 and 1. In case of r_{ij}, values can be anything between 0 and 1, including those numbers. When $S_i = 1$, the system is functioning properly. Malfunction is triggered when a state drops down to 0. Therefore models are commonly defined in inverse logic, according to which the zeros are good and the ones bad. Since our research is theoretical in this part, it is irrelevant whether we use positive or negative logic.

The actual values of probabilities vary for different real implemented component types. The r_{ij} values are greatly dependent on failure intensity, usually marked with λ. Another contribution to probability values is the interaction of states. The failure of a single state may cause other failures to occur. The intensity of a failure depends of many stochastic parameters, such as environmental parameters, technological parameters, quality of materials and manufacturing, etc. It depends on the factory and component quality.

Furthermore, the thermal characteristics of electronic devices are often under-designed, causing premature failure of electronic components, usually just after the expiry of the warranty period. One of the reasons we did not use actual experimental equipment data is such thermal design.

The weakness of thermal design does not lie in an important component, like the processor, but a component which is usually not even considered a computer part, i. e. the power converter. Since millions of electronic/electrical components cannot be analyzed in a clearly presentable way, only main parts are analyzed.

The reliability of the system can be further expressed if Table 5.1 is considered a reliability matrix, where r_{ij} are matrix members. Every column of the reliability matrix should be equal to 1. So, it can be written:

$$r_{00} + r_{F0} = 1 \tag{5.2}$$

$$r_{01} + r_{11} = 1 \tag{5.3}$$

$$r_{02} + r_{22} = 1 \tag{5.4}$$

$$r_{03} + r_{33} + r_{83} = 1 \tag{5.5}$$

$$r_{04} + r_{44} + r_{54} + r_{74} + r_{84} = 1 \qquad (5.6)$$

$$r_{05} + r_{15} + r_{55} + r_{85} = 1 \qquad (5.7)$$

$$r_{06} + r_{26} + r_{66} + r_{86} = 1 \qquad (5.8)$$

$$r_{07} + r_{77} + r_{87} = 1 \qquad (5.9)$$

$$r_{08} + r_{88} = 1 \qquad (5.10)$$

$$r_{1F} + r_{2F} + r_{3F} + r_{4F} + r_{5F} + r_{6F} + r_{7F} + r_{8F} + r_{FF} = 1 \qquad (5.11)$$

In order to link matrix members to an external datum, such as intensity of failure, λ, the inter-influences between states must be logically analyzed. Since system set up is such that λ_F cannot be directly achieved without inter-jumps to other states, it is by definition equal to 0 (because $r_{F0} = \lambda_F$, which is impossible and impossibility is expressed by zero). Therefore, we can write:

$$r_{00} = 1 - \lambda_F \Delta t = 1 \qquad (5.12)$$

The matrix member r_{01} is equal to failure intensity, which transits from state S_0 to S_1. It can be written:

$$r_{01} = \lambda_1 \Delta t \qquad (5.13)$$

In the same manner, we can see that one can write:

$$r_{11} = 1 - \lambda_1 \Delta t \qquad (5.14)$$

$$r_{02} = \lambda_2 \Delta t \qquad (5.15)$$

$$r_{22} = 1 - \lambda_2 \Delta t \qquad (5.16)$$

$$r_{03} = \lambda_3 \Delta t \qquad (5.17)$$

A complication occurs when analyzing r_{33}. This matrix member depends on failure intensity for S_3, but there is a possibility that failure 8 will interact with failure 3. If an interaction does not exist or is not possible, the corresponding $\alpha_{ij} = 0$. We marked these interactions with α_{ij}. We can write next few matrix members:

$$r_{33} = 1 - \lambda_3 \Delta t - \alpha_{83} \Delta t \qquad (5.18)$$

Matrix member r_{83} is about interaction of state 8 and 3, which can be described with α_{83}:

$$r_{83} = \alpha_{83}\Delta t \qquad (5.19)$$

Reliability matrix element r_{04} depend on intensity of failure for S_4:

$$r_{04} = \lambda_4\Delta t \qquad (5.20)$$

The diagonal matrix element for row 4 and column 4 is equal to:

$$r_{44} = 1 - (\lambda_4 + \alpha_{54} + \alpha_{74} + \alpha_{84})\Delta t \qquad (5.21)$$

Further reliability matrix elements for interaction of failures to S_4 are:

$$r_{54} = \alpha_{54}\Delta t \qquad (5.22)$$

$$r_{74} = \alpha_{74}\Delta t \qquad (5.23)$$

$$r_{84} = \alpha_{84}\Delta t \qquad (5.24)$$

From initial state, the system can degrade to state 5 with λ_5 intensity of failure:

$$r_{05} = \lambda_5\Delta t \qquad (5.25)$$

The member r_{15} can occur when failure 1 already happened and then failure 5 also occurs:

$$r_{15} = \alpha_{14}\Delta t \qquad (5.26)$$

The probability to stay in state 5 when the system was in state 5 for the next time period, Δt, can be then expressed with:

$$r_{55} = 1 - (\lambda_5 + \alpha_{15} + \alpha_{81})\Delta t \qquad (5.27)$$

Further, we can write down next matrix members:

$$r_{85} = \alpha_{85}\Delta t \qquad (5.28)$$

$$r_{06} = \lambda_6\Delta t \qquad (5.29)$$

$$r_{26} = \alpha_{26}\Delta t \qquad (5.30)$$

$$r_{66} = 1 - (\lambda_6 + \alpha_{26} + \alpha_{86})\Delta t \qquad (5.31)$$

$$r_{86} = \alpha_{86}\Delta t \qquad (5.32)$$

$$r_{07} = \lambda_7 \Delta t \tag{5.33}$$

$$r_{77} = 1 - (\lambda_7 + \alpha_{87})\Delta t \tag{5.34}$$

$$r_{87} = \alpha_{87} \Delta t \tag{5.35}$$

$$r_{08} = \lambda_8 \Delta t \tag{5.36}$$

$$r_{88} = 1 - \lambda_8 \Delta t \tag{5.37}$$

State of failure is further step to degradation of the system performance and can be expressed for all states with:

$$r_{1F} = (\lambda_1 + \alpha_{15})\Delta t \tag{5.38}$$

$$r_{2F} = (\lambda_2 + \alpha_{26})\Delta t \tag{5.39}$$

$$r_{3F} = \lambda_3 \Delta t \tag{5.40}$$

$$r_{4F} = \lambda_4 \Delta t \tag{5.41}$$

$$r_{5F} = (\lambda_5 + \alpha_{54})\Delta t \tag{5.42}$$

$$r_{6F} = \lambda_6 \Delta t \tag{5.43}$$

$$r_{7F} = (\alpha_{74} + \lambda_7)\Delta t \tag{5.44}$$

$$r_{8F} = (\lambda_8 + \alpha_{83} + \alpha_{84} + \alpha_{85} + \alpha_{86} + \alpha_{87})\Delta t \tag{5.45}$$

Finally, if the system is in the state of failure, it will remain in the faulty state in the next time instant, because we started the reliability analysis under such assumptions. However, the availability analysis will show that repairs make this impossible. So, for reliability analysis, we can consider that:

$$r_{FF} = 1 \tag{5.46}$$

It should be noted that α is an attempt to compensate for the fact that this is a relatively simple model, which includes only 0 and 1 for states. If multiple possibilities for states were allowed, the α-s could be skipped. The usual reliability analysis leads to the triangle matrix, which is less complex than the one we analyzed.

Table 5.1 served as a source for the mathematical expression of the problem. Using columns, for the first one, it can be written:

$$P_{S0}(t + \Delta t) = r_{00}P_{S0}(t) + r_{F0}P_{SF}(t) \tag{5.47}$$

which means that a system can be in the operating state functioning correctly in the next time instant with probability $P_{S0}(t + \Delta t)$. This probability is equal to the probability of remaining in state 0 and returning from fault to operating state. For non-reparable systems the second component is equal to zero.

The probability that a system operates with unsatisfactory camera performance at a given moment $t + \Delta t$ is equal to the probability that the system would change state to S_1 from S_0 if it was in S_0:

$$P_{S1}(t + \Delta t) = r_{01}P_{S0}(t) + r_{11}P_{S1}(t) \tag{5.48}$$

Environmental characteristics can cause failure if the system was operating properly (S_0) until something unfavorable occurred:

$$P_{S2}(t + \Delta t) = r_{22}P_{S2}(t) + r_{01}P_{S0}(t) \tag{5.49}$$

The system jumps to S_3 if S_8 occurs or directly from S_0. However, the system can stay in S_3 if it was in S_3 in the preceding time interval:

$$P_{S3}(t + \Delta t) = r_{03}P_{S0}(t) + r_{33}P_{S3}(t) + r_{83}P_{S8}(t) \tag{5.50}$$

The system may jump to S_4 in the next time interval if it was operating properly (S_0), or was already in S_4 or if S_5, S_7 or S_8 occurred:

$$\begin{aligned} P_{S4}(t + \Delta t) = r_{04}P_{S0}(t) + r_{44}P_{S4}(t) \\ + r_{54}P_{S5}(t) + r_{74}P_{S7}(t) + r_{84}P_{S8}(t) \end{aligned} \tag{5.51}$$

It is possible that the system jumps to S_5 in the next time interval if it was operating properly (S_0), or it was already in S_5 or if S_1 or S_8 occurred:

$$P_{S5}(t + \Delta t) = r_{05}P_{S0}(t) + r_{55}P_{S5}(t) + r_{15}P_{S1}(t) + r_{85}P_{S8}(t) \tag{5.52}$$

It is possible that the system jumps to S_6 in the next time interval if it was operating properly (S_0), or it was already in S_6 or if S_2 or S_8 occurred:

$$P_{S6}(t + \Delta t) = r_{06}P_{S0}(t) + r_{66}P_{S6}(t) + r_{26}P_{S2}(t) + r_{86}P_{S8}(t) \tag{5.53}$$

It is possible that the system jumps to S_7 in the next time interval if it was operating properly (S_0), or it was already in S_7 or if S_8 occurred:

$$P_{S7}(t + \Delta t) = r_{07}P_{S0}(t) + r_{77}P_{S7}(t) + r_{87}P_{S8}(t) \tag{5.54}$$

However, the system can be in S_8 only if it was in S_0 before:

$$P_{S8}(t + \Delta t) = r_{08}P_{S0}(t) + r_{88}P_{S8}(t) \tag{5.55}$$

The system can jump to failure state (S_F) if any fault occurred in previous time interval:

$$
\begin{aligned}
P_{SF}(t + \Delta t) = {} & r_{1F}P_{S1}(t) + r_{2F}P_{S2}(t) + r_{3F}P_{S3}(t) + r_{4F}P_{S4}(t) \\
& + r_{5F}P_{S5}(t) + r_{6F}P_{S6}(t) + r_{7F}P_{S7}(t) + r_{8F}P_{S8}(t) + r_{FF}P_{SF}(t)
\end{aligned}
\tag{5.56}
$$

Assuming that $r_{00} = 1$ and $r_{0F} = 0$, because a part of the system has to malfunction to cause the failure of the entire system, we have:

$$P_{S0}(t + \Delta t) = 1 \cdot P_{S0}(t) + 0 \cdot P_{SF}(t) = P_{S0}(t) \tag{5.57}$$

Rearranging expression (5.56) for P_{SF}, we obtain:

$$P_{S1}(t + \Delta t) = \lambda_1 \Delta t P_{S0}(t) + (1 - \lambda_1 \Delta t)P_{S1}(t) \tag{5.58}$$

Rearranging expression (5.49) for P_{S2}, we obtain:

$$P_{S2}(t + \Delta t) = (1 - \lambda_2 \Delta t)P_{S2}(t) + \lambda_1 \Delta t P_{S0}(t) \tag{5.59}$$

Rearranging expression (5.50) for P_{S3}, we obtain:

$$P_{S3}(t + \Delta t) = \lambda_3 \Delta t P_{S0}(t) + (1 - \lambda_3 \Delta t - \alpha_{83} \Delta t)P_{S3}(t) + \alpha_{83} \Delta t P_{S8}(t) \tag{5.60}$$

Rearranging expression (5.51) for P_{S4}, we obtain:

$$
\begin{aligned}
P_{S4}(t + \Delta t) = {} & \lambda_4 \Delta t P_{S0}(t) + [1 - (\lambda_4 + \alpha_{54} + \alpha_{74} + \alpha_{84})\Delta t]P_{S4}(t) \\
& + \alpha_{54} \Delta t P_{S5}(t) + \alpha_{74} \Delta t P_{S7}(t) + \alpha_{84} \Delta t P_{S8}(t)
\end{aligned}
\tag{5.61}
$$

Rearranging expression (5.52) for P_{S5}, we obtain:

$$
\begin{aligned}
P_{S5}(t + \Delta t) = {} & \lambda_5 \Delta t P_{S0}(t) + [1 - (\lambda_5 + \alpha_{15} + \alpha_{81})\Delta t]P_{S5}(t) \\
& + \alpha_{14} \Delta t P_{S1}(t) + \alpha_{85} \Delta t P_{S8}(t)
\end{aligned}
\tag{5.62}
$$

Rearranging expression (5.53) for P_{S6}, we obtain:

$$
\begin{aligned}
P_{S6}(t + \Delta t) = {} & \lambda_6 \Delta t P_{S0}(t) + [1 - (\lambda_6 + \alpha_{26} + \alpha_{86})\Delta t]P_{S6}(t) \\
& + \alpha_{26} \Delta t P_{S2}(t) + \alpha_{86} \Delta t P_{S8}(t)
\end{aligned}
\tag{5.63}
$$

Rearranging expression (5.54) for P_{S7}, we obtain:

$$P_{S7}(t + \Delta t) = \lambda_7 \Delta t P_{S0}(t) + [1 - (\lambda_7 + \alpha_{87})\Delta t]P_{S7}(t) + \alpha_{87}\Delta t P_{S8}(t) \qquad (5.64)$$

Rearranging expression (5.55) for P_{S8}, we obtain:

$$P_{S8}(t + \Delta t) = \lambda_8 \Delta t P_{S0}(t) + [1 - \lambda_8 \Delta t]P_{S8}(t) \qquad (5.65)$$

Rearranging expression (5.56) for P_{SF}, we obtain:

$$
\begin{aligned}
P_{SF}(t + \Delta t) = {} & (\lambda_1 + \alpha_{15})\Delta t P_{S1}(t) + (\lambda_2 + \alpha_{26})\Delta t P_{S2}(t) \\
& + \lambda_3 \Delta t P_{S3}(t) + \lambda_4 \Delta t P_{S4}(t) \\
& + (\lambda_5 + \alpha_{54})\Delta t P_{S5}(t) + \lambda_6 \Delta t P_{S6}(t) + (\alpha_{74} + \lambda_7)\Delta t P_{S7}(t) \\
& + (\lambda_8 + \alpha_{83} + \alpha_{84} + \alpha_{85} + \alpha_{86} + \alpha_{87})\Delta t P_{S8}(t) + P_{SF}(t)
\end{aligned}
\qquad (5.66)
$$

If we set a limit for the translation of equations from discrete time to continued functions, we can derive differential equation for P_{S0}:

$$\lim_{\Delta t \to 0} \frac{P_{S0}(t + \Delta t) - P_{S0}(t)}{\Delta t} = 0 \qquad (5.67)$$

$$\frac{dP_{S0}(t)}{dt} = 0 \qquad (5.68)$$

Taking account initial conditions, we can calculate integration constant:

$$P_{S0} = const. = 1 \qquad (5.69)$$

Similarly, we can derive differential equation for S_1:

$$\lim_{\Delta t \to 0} \frac{P_{S1}(t + \Delta t) - P_{S1}(t)}{\Delta t} = \lim_{\Delta t \to 0} [\lambda_1 P_{S0}(t) + (-\lambda_1)P_{S1}(t)] \qquad (5.70)$$

$$\frac{dP_{S1}(t)}{dt} = \lambda_1 P_{S0}(t) - \lambda_1 P_{S1}(t) \qquad (5.71)$$

or rearranging to recognized form:

$$\frac{dP_{S1}(t)}{dt} + \lambda_1 P_{S1}(t) = \lambda_1 P_{S0}(t) \qquad (5.72)$$

Putting $P_{S0} = 1$, we have:

$$\frac{dP_{S1}(t)}{dt} + \lambda_1 P_{S1}(t) = \lambda_1 \qquad (5.73)$$

This equation leads to general solution:

$$P_{S1}(t) = C_1 e^{-\lambda_1 t} + \lambda_1 t e^{-\lambda_1 t} \tag{5.74}$$

With assumption that $P_{S1}(0) = 0$, $C_1 = 0$ and we have final solution:

$$P_{S1}(t) = \lambda_1 \cdot t \cdot e^{-\lambda_1 \cdot t} \tag{5.75}$$

Similar derivation can be performed for other states. For S_2, we obtain:

$$\lim_{\Delta t \to 0} \frac{P_{S2}(t + \Delta t) - P_{S2}(t)}{\Delta t} = \lim_{\Delta t \to 0} [(-\lambda_2) P_{S2}(t) + \lambda_1 P_{S0}(t)] \tag{5.76}$$

$$\frac{dP_{S2}(t)}{dt} = -\lambda_2 P_{S2}(t) + \lambda_1 P_{S0}(t) \tag{5.77}$$

Since P_{S0} is assumed is equal to 1, we can simplify equation:

$$\frac{dP_{S2}(t)}{dt} + \lambda_2 P_{S2}(t) = \lambda_1 P_{S0} = \lambda_1 \tag{5.78}$$

which leads to general solution:

$$P_{S2}(t) = C_2 e^{-\lambda_2 t} + \lambda_2 t e^{-\lambda_2 t} \tag{5.79}$$

With assumption that $P_{S2}(0) = 0$, $C_2 = 0$ and we have final solution:

$$P_{S2}(t) = \lambda_2 \cdot t \cdot e^{-\lambda_2 \cdot t} \tag{5.80}$$

Similarly, the derivation for S_8 can be performed:

$$\lim_{\Delta t \to 0} \frac{P_{S8}(t + \Delta t) - P_{S8}(t)}{\Delta t} = \lim_{\Delta t \to 0} [\lambda_8 P_{S0}(t) + [-\lambda_8] P_{S8}(t)] \tag{5.81}$$

$$\frac{dP_{S8}(t)}{dt} = \lambda_8 P_{S0}(t) - \lambda_8 P_{S8}(t) \tag{5.82}$$

$$\frac{dP_{S8}(t)}{dt} + \lambda_8 P_{S8}(t) = \lambda_8 \tag{5.83}$$

and the solution is:

$$P_{S8}(t) = \lambda_8 \cdot t \cdot e^{-\lambda_8 \cdot t} \tag{5.84}$$

Further states have a bit more difficult derivation and solution. The complete procedure in Matlab is explained in Appendix A.1. Here are some basic steps and solutions. The differential equation for S_3 involves two unknown states, one of which was previously calculated and the solution can be used here:

$$\frac{dP_{S3}(t)}{dt} + (\lambda_3 + \alpha_{83})P_{S3}(t) = \lambda_3 1 + \alpha_{83}P_{S8}(t) \tag{5.85}$$

The solution obtained by Matlab is:

$$
\begin{aligned}
P_{S3} = &\left(\frac{\lambda_3 e^{(\alpha_{83}+\lambda_3)t}}{\alpha_{83}+\lambda_3} - \frac{\alpha_{83}\lambda_8 e^{(\alpha_{83}+\lambda_3-\lambda_8)t}}{(\alpha_{83}+\lambda_3-\lambda_8)^2} + \frac{\alpha_{83}\lambda_8 t e^{(\alpha_{83}+\lambda_3-\lambda_8)t}}{\alpha_{83}+\lambda_3-\lambda_8} \right) \cdot e^{-t(\alpha_{83}+\lambda_3)} \\
&- \left(\frac{\lambda_3}{\alpha_{83}+\lambda_3} - \frac{\alpha_{83}\lambda_8}{(\alpha_{83}+\lambda_3-\lambda_8)^2} \right) e^{-t(\alpha_{83}+\lambda_3)}
\end{aligned}
\tag{5.86}
$$

By deriving equation for S_4, we obtain:

$$
\begin{aligned}
\frac{dP_{S4}(t)}{dt} + (\lambda_4 &+ \alpha_{54} + \alpha_{74} + \alpha_{84})P_{S4}(t) \\
&= \lambda_4 + \alpha_{54}P_{S5}(t) + \alpha_{74}P_{S7}(t) + \alpha_{84}P_{S8}(t)
\end{aligned}
\tag{5.87}
$$

and the solution is:

$$
\begin{aligned}
P_{S4} = &\left(\frac{(\lambda_4 + \alpha_{54} \cdot P_{S5})e^{(\alpha_{54}+\alpha_{74}+\alpha_{84}+\lambda_4)t}}{\alpha_{54}+\alpha_{74}+\alpha_{84}+\lambda_4} \right) \\
&- \frac{\alpha_{74}\lambda_8 e^{(\alpha_{54}+\alpha_{74}+\alpha_{84}+\lambda_4-\lambda_8)t}}{(\alpha_{54}+\alpha_{74}+\alpha_{84}+\lambda_4-\lambda_8)^2} \\
&+ \frac{\alpha_{74}\lambda_8 t e^{(\alpha_{54}+\alpha_{74}+\alpha_{84}+\lambda_4-\lambda_8)t}}{\alpha_{54}+\alpha_{74}+\alpha_{84}+\lambda_4-\lambda_8} e^{-t(\alpha_{54}+\alpha_{74}+\alpha_{84}+\lambda_4)} \\
&- \left(\frac{\lambda_4 + \alpha_{54}P_{S5}}{\alpha_{54}+\alpha_{74}+\alpha_{84}+\lambda_4} - \frac{\alpha_{74}\lambda_8}{(\alpha_{54}+\alpha_{74}+\alpha_{84}+\lambda_4-\lambda_8)^2} \right) \\
&\times e^{-(\alpha_{54}+\alpha_{74}+\alpha_{84}+\lambda_4)t}
\end{aligned}
\tag{5.88}
$$

The equation for S_5 has three unknowns, but two were previously calculated and their solutions can be used in this equation.

$$\frac{dP_{S5}(t)}{dt} + (\lambda_5 + \alpha_{15} + \alpha_{81})P_{S5}(t) = \lambda_5 + \alpha_{14}P_{S1}(t) + \alpha_{85}P_{S8}(t) \tag{5.89}$$

If we use initial conditions, solution of the differential equation is:

$$P_{S5} = \left(\frac{\alpha_{14}\lambda_1}{(\alpha_{15} + \alpha_{81} - \lambda_1 + \lambda_2)^2} - \frac{\lambda_5}{\alpha_{15} + \alpha_{81} + \lambda_2} \right.$$

$$\left. + \frac{\alpha_{85}\lambda_8}{(\alpha_{15} + \alpha_{81} + \lambda_2 - \lambda_8)^2} \right) e^{-t(\alpha_{15}+\alpha_{81}+\lambda_2)}$$

$$+ \left(\frac{\lambda_5 e^{\alpha_{15}t+\alpha_{81}t+\lambda_2 t}}{\alpha_{15} + \alpha_{81} + \lambda_2} - \frac{\alpha_{14}\lambda_1 e^{\alpha_{15}t+\alpha_{81}t-\lambda_1 t+\lambda_2 t}}{(\alpha_{15} + \alpha_{81} - \lambda_1 + \lambda_2)^2} \right. \tag{5.90}$$

$$- \frac{\alpha_{85}\lambda_8 e^{\alpha_{15}t+\alpha_{81}t+\lambda_2 t-\lambda_8 t}}{(\alpha_{15} + \alpha_{81} + \lambda_2 - \lambda_8)^2} + \frac{\alpha_{14}\lambda_1 t e^{\alpha_{15}t+\alpha_{81}t-\lambda_1 t+\lambda_2 t}}{\alpha_{15} + \alpha_{81} - \lambda_1 + \lambda_2}$$

$$\left. + \frac{\alpha_{85}\lambda_8 t e^{\alpha_{15}t+\alpha_{81}t+\lambda_2 t-\lambda_8 t}}{\alpha_{15} + \alpha_{81} + \lambda_2 - \lambda_8} \right) e^{-t(\alpha_{15}+\alpha_{81}+\lambda_2)}$$

Equation for P_{S6} also has three unknowns, and two of them are calculated before:

$$\frac{dP_{S6}(t)}{dt} + (\lambda_6 + \alpha_{26} + \alpha_{86})P_{S6}(t) = \lambda_6 + \alpha_{26}P_{S2}(t) + \alpha_{86}P_{S8}(t) \tag{5.91}$$

Including initial conditions, we obtain final solution:

$$P_{S6} = \left(\frac{\lambda_6 e^{\alpha_{26}t+\alpha_{86}t+\lambda_6 t}}{\alpha_{26} + \alpha_{86} + \lambda_6} + \frac{\alpha_{26}\lambda_2 e^{\alpha_{26}t+\alpha_{86}t-\lambda_2 t+\lambda_6 t}}{(\alpha_{26} + \alpha_{86} - \lambda_2 + \lambda_6)^2} \right.$$

$$- \frac{\alpha_{86}\lambda_8 e^{(\alpha_{26}+\alpha_{86}+\lambda_6-\lambda_8)t}}{(\alpha_{26} + \alpha_{86} + \lambda_6 - \lambda_8)^2} - \frac{\alpha_{26}\lambda_2 t e^{\alpha_{26}t+\alpha_{86}t-\lambda_2 t+\lambda_6 t}}{\alpha_{26} + \alpha_{86} - \lambda_2 + \lambda_6}$$

$$\left. + \frac{\alpha_{86}\lambda_8 t e^{\alpha_{26}t+\alpha_{86}t+\lambda_6 t-\lambda_8 t}}{\alpha_{26} + \alpha_{86} + \lambda_6 - \lambda_8} \right) e^{-t(\alpha_{26}+\alpha_{86}-\lambda_6)} \tag{5.92}$$

$$- \left(\frac{\lambda_6}{\alpha_{26} + \alpha_{86} + \lambda_6} + \frac{\alpha_{26}\lambda_2}{(\alpha_{26} + \alpha_{86} + \lambda_6 - \lambda_2)^2} \right.$$

$$\left. - \frac{\alpha_{86}\lambda_8}{(\alpha_{26} + \alpha_{86} + \lambda_6 - \lambda_8)^2} \right) e^{-t(\alpha_{26}+\alpha_{86}+\lambda_6)}$$

Similar procedure is performed in order to obtain solution for S_7:

$$\lim_{\Delta t \to 0} \frac{P_{S7}(t + \Delta t) - P_{S7}(t)}{\Delta t} = \lim_{\Delta t \to 0} \{ \lambda_7 P_{S0}(t)$$

$$+ [-(\lambda_7 + \alpha_{87})]P_{S7}(t) + \alpha_{87}P_{S8}(t) \} \tag{5.93}$$

$$\frac{dP_{S7}(t)}{dt} = \lambda_7 P_{S0}(t) - (\lambda_7 + \alpha_{87})P_{S7}(t) + \alpha_{87}P_{S8}(t) \tag{5.94}$$

Rearranging, we have:

$$\frac{dP_{S7}(t)}{dt} + (\lambda_7 + \alpha_{87})P_{S7}(t) = \lambda_7 + \alpha_{87}P_{S8}(t) \tag{5.95}$$

General solution for P_{S7} is:

$$
\begin{aligned}
P_{S7} &= C_2 e^{-(\lambda_7 + \alpha_{87})t} + \frac{\frac{\lambda_7 \cdot e^{(\lambda_7 + \alpha_{87})t}}{\lambda_7 + \alpha_{87}} - \alpha_{87} \cdot \lambda_8 e^{(\lambda_7 - \lambda_8 + \alpha_{87})t}}{(\lambda_7 + \alpha_{87} - \lambda_8)^2} \\
&\quad + \frac{\alpha_{87} \cdot \lambda_8 t e^{(\lambda_7 - \lambda_8 + \alpha_{87})t}}{(\lambda_7 + \alpha_{87}) - \lambda_8} e^{-(\lambda_7 + \alpha_{87})t} \\
&= C_2 e^{-(\lambda_7 + \alpha_{87})t} + \frac{\frac{\lambda_7 \cdot e^{(\lambda_7 + \alpha_{87})t}}{\lambda_7 + \alpha_{87}} - \alpha_{87} \cdot \lambda_8 e^{(\lambda_7 - \lambda_8 + \alpha_{87})t}}{(\lambda_7 + \alpha_{87} - \lambda_8)^2} \\
&\quad + \frac{\alpha_{87} \cdot \lambda_8 t e^{-\lambda_8 t}}{\lambda_7 + \alpha_{87} - \lambda_8}
\end{aligned}
\tag{5.96}
$$

Including initial conditions, we obtain:

$$
\begin{aligned}
P_{S7} &= \left(\frac{\lambda_7 e^{(\lambda_7 + \alpha_{87})t}}{\lambda_7 + \alpha_{87}} - \frac{\alpha_{87} \lambda_8 e^{(\lambda_7 + \alpha_{87} - \lambda_8)t}}{(\lambda_7 + \alpha_{87} - \lambda_8)^2} \right. \\
&\quad \left. + \frac{\alpha_{87} \lambda_8 t e^{((\lambda_7 + \alpha_{87} - \lambda_8)t)}}{\lambda_7 + \alpha_{87} - \lambda_8} \right) e^{-(\lambda_7 + \alpha_{87})t} \\
&\quad - \left(\frac{\lambda_7}{\lambda_7 + \alpha_{87}} - \frac{\alpha_{87} \lambda_8}{(\lambda_7 + \alpha_{87} - \lambda_8)^2} \right) e^{(\lambda_7 + \alpha_{87})t}
\end{aligned}
\tag{5.97}
$$

According to solution procedure in the Appendix A.1, the solution for failure state is:

$$
\begin{aligned}
P_{SF} = t \cdot (&\alpha_{15} \cdot P_{S1} + \alpha_{26} \cdot P_{S2} + \alpha_{54} \cdot P_{S5} + \alpha_{74} \cdot P_{S7} \\
&+ (\alpha_{83} + \alpha_{84} + \alpha_{85} + \alpha_{86} + \alpha_{87})P_{S8} \\
&+ \lambda_1 P_{S1} + \lambda_2 P_{S2} + \lambda_3 P_{S3} + \lambda_4 P_{S4} \\
&+ \lambda_5 P_{S5} + \lambda_6 P_{S6} + \lambda_7 P_{S7} + \lambda_8 P_{S8})
\end{aligned}
\tag{5.98}
$$

If we want numbers, we have to obtain the real values either by experiments or from the manufacturer and use them.

5.2 Availability Analysis

Since when dealing with availability, the failure rate and repair rate must be taken into consideration, equations for availability are more complex than those for reliability. The system is unavailable when it fails until it is repaired.

Figure 5.2 shows a model for the proposed system's availability. Since visual quality control system is a reparable system, the system can return from state of fault to operating state. However, if the system changes its state form S_0 to S_1, it is automatically in state S_F. Therefore, it is not possible to return to S_0 from S_1, but rather to S_0 from S_F. Therefore $w_{10} = 0$.

Table 5.2 illustrates transitions between states for the next time period. The initial states are in the first column, $P_S(t)$, and the following states in columns P_{S0} to P_{SF} with common designation $P_S(t + \Delta t)$. Possible transitions are expressed with probabilities, a_{ij}, and impossible transitions with zero. When dealing with availability we should bear in mind that there are two ways to get from failure state to P_{S0}. First, the failure can be repaired "on the spot", which means that when a part fails, it is repaired without failure of the entire system (path: $S_i - S_0$). Second, a failure can escalate from failure of one subsystem to the failure of the entire system, in which case first the malfunctioning component is repaired and then the entire system, meaning that the repair path is $S_F - S_i - S_0$. Furthermore, it should be noted that while some subsystems can actually be repaired "on the spot", others cannot. So, we can say that the intensity of repairs, μ_i, depends on the relationship between the previous states and the state of failure. We naturally want to express theoretical options aij with known intensities of repairs, μ_i, and intensities of failure, λ_i.

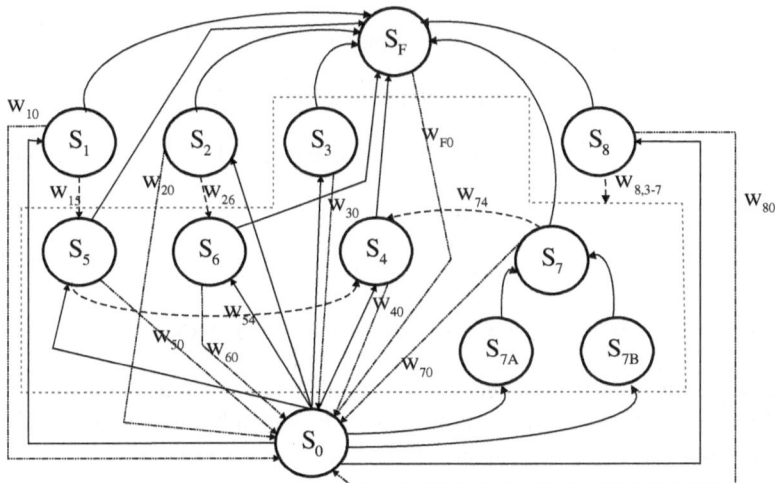

Fig. 5.2 System availability model

Table 5.2 State transitions for availability

$P_S(t)$	$P_S(t+\Delta t)$									
	P_{S0}	P_{S1}	P_{S2}	P_{S3}	P_{S4}	P_{S5}	P_{S6}	P_{S7}	P_{S8}	P_{SF}
P_{S0}	a_{00}	a_{01}	a_{02}	a_{03}	a_{04}	a_{05}	a_{06}	a_{07}	a_{08}	0
P_{S1}	a_{10}	a_{11}	0	0	0	a_{15}	0	0	0	a_{1F}
P_{S2}	a_{20}	0	a_{22}	0	0	0	a_{26}	0	0	a_{2F}
P_{S3}	a_{30}	0	0	a_{33}	0	0	0	0	0	a_{3F}
P_{S4}	a_{40}	0	0	0	a_{44}	0	0	0	0	a_{4F}
P_{S5}	a_{50}	0	0	0	a_{54}	a_{55}	0	0	0	a_{5F}
P_{S6}	a_{60}	0	0	0	0	0	a_{66}	0	0	a_{6F}
P_{S7}	a_{70}	0	0	0	a_{74}	0	0	a_{77}	0	a_{7F}
P_{S8}	a_{80}	0	0	a_{83}	a_{84}	a_{85}	a_{86}	a_{87}	a_{88}	a_{8F}
P_{SF}	a_{F0}	a_{F1}	a_{F2}	a_{F3}	a_{F4}	a_{F5}	a_{F6}	a_{F7}	a_{F8}	a_{FF}

In order to identify a solution for this system of 10 equations and 10 unknowns, we will derive differential equations and try to find solutions for the unknowns. First, we will explore the fact that columns should be equal to one. Therefore we can write:

$$a_{00} + a_{10} + a_{20} + a_{30} + a_{40} + a_{50} + a_{60} + a_{70} + a_{80} + a_{F0} = 1 \tag{5.99}$$

$$a_{01} + a_{11} + a_{F1} = 1 \tag{5.100}$$

$$a_{02} + a_{22} + a_{F2} = 1 \tag{5.101}$$

$$a_{03} + a_{33} + a_{83} + a_{F3} = 1 \tag{5.102}$$

$$a_{04} + a_{44} + a_{54} + a_{74} + a_{84} + a_{F4} = 1 \tag{5.103}$$

$$a_{05} + a_{15} + a_{55} + a_{85} + a_{F5} = 1 \tag{5.104}$$

$$a_{06} + a_{26} + a_{66} + a_{86} + a_{F6} = 1 \tag{5.105}$$

$$a_{07} + a_{77} + a_{87} + a_{F7} = 1 \tag{5.106}$$

$$a_{08} + a_{88} + a_{F8} = 1 \tag{5.107}$$

$$a_{1F} + a_{2F} + a_{3F} + a_{4F} + a_{5F} + a_{6F} + a_{7F} + a_{8F} + a_{FF} = 1 \tag{5.108}$$

In order to link matrix members to an external datum, such as intensity of failure, λ_i, and intensity of repairs, μ_i, it is necessary to logically analyze inter-influences between states. The system set up is such that $\lambda_F = 0$, because one subsystem has to fail in order for the system to fail. However, μ_F is possible because the system is reparable. Furthermore, a_{10} to a_{80} are dependent on λ_1 to λ_8. Therefore, we can write:

$$a_{00} = 1 - \sum_{i=1}^{8} a_{i0} - a_{F0} \tag{5.109}$$
$$= 1 - (\lambda_1 + \lambda_2 + \lambda_3 + \lambda_4 + \lambda_5 + \lambda_6 + \lambda_7 + \lambda_8)\Delta t$$

$$a_{10} = \lambda_1 \Delta t \tag{5.110}$$

$$a_{20} = \lambda_2 \Delta t \tag{5.111}$$

$$a_{30} = \lambda_3 \Delta t \tag{5.112}$$

$$a_{40} = \lambda_4 \Delta t \tag{5.113}$$

$$a_{50} = \lambda_5 \Delta t \tag{5.114}$$

$$a_{60} = \lambda_6 \Delta t \tag{5.115}$$

$$a_{70} = \lambda_7 \Delta t \tag{5.116}$$

$$a_{80} = \lambda_8 \Delta t \tag{5.117}$$

$$a_{F0} = 0 \tag{5.118}$$

Availabilities in transition from fault to other states except full-functioning, S_0, are:

$$a_{F1} = \lambda_1 \Delta t \tag{5.119}$$

$$a_{F2} = \lambda_2 \Delta t \tag{5.120}$$

$$a_{F3} = \lambda_3 \Delta t \tag{5.121}$$

$$a_{F4} = \lambda_4 \Delta t \tag{5.122}$$

$$a_{F5} = \lambda_5 \Delta t \tag{5.123}$$

$$a_{F6} = \lambda_6 \Delta t \tag{5.124}$$

$$a_{F7} = \lambda_7 \Delta t \tag{5.125}$$

$$a_{F8} = \lambda_8 \Delta t \tag{5.126}$$

The probability that a failed system will remain in fault state in the next time period is equal to the probability that it will not be repaired in the current time period and the probability that it is in the fault state, which is 100 % minus all availabilities a_{iF}:

$$a_{FF} = 1 - \sum_{j=1}^{8} a_{iF}$$
$$= 1 - \Delta t(\mu_1 + \alpha_{15} + \mu_2 + \alpha_{26} + \mu_3 + \mu_4 + \mu_5 + \alpha_{54} + \mu_6$$
$$+ \mu_7 + \alpha_{74} + \mu_8 + \alpha_{83} + \alpha_{84} + \alpha_{85} + \alpha_{86} + \alpha_{87})$$
$$(5.127)$$

Probability to get into S_2 from the all-functioning state, S_0, is directly dependent on μ_2:

$$a_{02} = \mu_2 \Delta t \qquad (5.128)$$

All diagonal elements are equal to 1 minus other non-zero members of the column. Therefore, element a_{22} is:

$$a_{22} = 1 - \lambda_2 \Delta t - \mu_2 \Delta t \qquad (5.129)$$

$$a_{11} = 1 - \lambda_1 \Delta t - \mu_1 \Delta t \qquad (5.130)$$

$$a_{33} = 1 - \lambda_3 \Delta t - \alpha_{83} \Delta t - \mu_3 \Delta t \qquad (5.131)$$

$$a_{44} = 1 - (\lambda_4 + \mu_4 + \alpha_{54} + \alpha_{74} + \alpha_{84}) \Delta t \qquad (5.132)$$

$$a_{55} = 1 - (\lambda_5 + \mu_5 + \alpha_{15} + \alpha_{81}) \Delta t \qquad (5.133)$$

$$a_{66} = 1 - (\lambda_6 + \alpha_{26} + \alpha_{86} + \mu_6) \Delta t \qquad (5.134)$$

$$a_{77} = 1 - (\lambda_7 + \mu_7 + \alpha_{87}) \Delta t \qquad (5.135)$$

$$a_{88} = 1 - (\mu_8 + \lambda_8) \Delta t \qquad (5.136)$$

Availabilities a_{0i} can be expressed through intensities of repairs:

$$a_{01} = \mu_1 \Delta t \qquad (5.137)$$

$$a_{03} = \mu_3 \Delta t \qquad (5.138)$$

$$a_{04} = \mu_4 \Delta t \qquad (5.139)$$

$$a_{05} = \mu_5 \Delta t \qquad (5.140)$$

$$a_{06} = \mu_6 \Delta t \qquad (5.141)$$

$$a_{07} = \mu_7 \Delta t \qquad (5.142)$$

$$a_{08} = \mu_8 \Delta t \qquad (5.143)$$

The intermixed states are more difficult to express and there are no generalizations for them. Since availability depends on several relationships, it is expressed by the intermixed parameter, α_{ij}. Furthermore, the intensities of repairs are not linked to mixed states, because there are no connections in the model for availability (see Fig. 5.2). So a_{83} and a_{54} are:

$$a_{83} = \alpha_{83}\Delta t \tag{5.144}$$

$$a_{54} = \alpha_{54}\Delta t \tag{5.145}$$

In theory, i.e. a_{83} should include μ_8 and should be written as:

$$a_{83} = \alpha_{83}\Delta t + \mu_8\Delta t \tag{5.146}$$

But this is not the case. Similar consideration is valid for a_{54}. Other mixed states are:

$$a_{74} = \alpha_{74}\Delta t \tag{5.147}$$

$$a_{84} = \alpha_{84}\Delta t \tag{5.148}$$

$$a_{15} = \alpha_{15}\Delta t \tag{5.149}$$

$$a_{85} = \alpha_{85}\Delta t \tag{5.150}$$

$$a_{26} = \alpha_{26}\Delta t \tag{5.151}$$

$$a_{86} = \alpha_{86}\Delta t \tag{5.152}$$

$$a_{87} = \alpha_{87}\Delta t \tag{5.153}$$

Probabilities to get into the failure state from some state other then 0 are expressed with:

$$a_{1F} = \mu_1\Delta t + \alpha_{15}\Delta t \tag{5.154}$$

$$a_{2F} = \mu_2\Delta t + \alpha_{26}\Delta t \tag{5.155}$$

$$a_{3F} = \mu_3\Delta t \tag{5.156}$$

$$a_{4F} = \mu_4\Delta t \tag{5.157}$$

$$a_{5F} = \mu_5\Delta t + \alpha_{54}\Delta t \tag{5.158}$$

$$a_{6F} = \mu_6 \Delta t \tag{5.159}$$

$$a_{7F} = \mu_7 \Delta t + \alpha_{74} \Delta t \tag{5.160}$$

$$a_{8F} = \mu_8 \Delta t + (\alpha_{83} + \alpha_{84} + \alpha_{85} + \alpha_{86} + \alpha_{87}) \Delta t \tag{5.161}$$

Based on previous expressions for availabilities, we can write probability functions. The probability of being in a no-failure state is equal to the probability of having been in such a state in the previous time period and having been repaired in the meantime:

$$
\begin{aligned}
P_{S0}(t + \Delta t) = {} & a_{00} P_{S0}(t) + a_{10} P_{S1}(t) + a_{20} P_{S2}(t) \\
& + a_{30} P_{S3}(t) + a_{40} P_{S4}(t) + a_{50} P_{S5}(t) \\
& + a_{60} P_{S6}(t) + a_{70} P_{S7}(t) + a_{80} P_{S8}(t) + a_{F0} P_{SF}(t)
\end{aligned}
\tag{5.162}
$$

Substituting availabilities with intensities of repairs and failure, we obtain:

$$
\begin{aligned}
P_{S0}(t + \Delta t) = {} & [1 - (\lambda_1 + \lambda_2 + \lambda_3 + \lambda_4 + \lambda_5 + \lambda_6 + \lambda_7 + \lambda_8)\Delta t] P_{S0}(t) \\
& + \lambda_1 \Delta t P_{S1}(t) + \lambda_2 \Delta t P_{S2}(t) + \lambda_3 \Delta t P_{S3}(t) + \lambda_4 \Delta t P_{S4}(t) + \lambda_5 \Delta t P_{S5}(t) \\
& + \lambda_6 \Delta t P_{S6}(t) + \lambda_7 \Delta t P_{S7}(t) + \lambda_8 \Delta t P_{S8}(t) + 0 \cdot P_{SF}(t)
\end{aligned}
\tag{5.163}
$$

and:

$$
\begin{aligned}
P_{S0}(t + \Delta t) = {} & [1 - (\lambda_1 + \lambda_2 + \lambda_3 + \lambda_4 + \lambda_5 + \lambda_6 + \lambda_7 + \lambda_8)\Delta t] P_{S0}(t) \\
& + \lambda_1 \Delta t P_{S1}(t) + \lambda_2 \Delta t P_{S2}(t) + \lambda_3 \Delta t P_{S3}(t) + \lambda_4 \Delta t P_{S4}(t) + \lambda_5 \Delta t P_{S5}(t) \\
& + \lambda_6 \Delta t P_{S6}(t) + \lambda_7 \Delta t P_{S7}(t) + \lambda_8 \Delta t P_{S8}(t)
\end{aligned}
\tag{5.164}
$$

Putting limit to zero:

$$
\begin{aligned}
\lim_{\Delta t \to 0} \frac{P_{S0}(t + \Delta t) - P_{S0}(t)}{\Delta t} = {} & \lim_{\Delta t \to 0} [-(\lambda_1 + \lambda_2 + \lambda_3 + \lambda_4 + \lambda_5 + \lambda_6 + \lambda_7 + \lambda_8) P_{S0}(t) \\
& + \lambda_1 P_{S1}(t) + \lambda_2 P_{S2}(t) + \lambda_3 P_{S3}(t) + \lambda_4 P_{S4}(t) \\
& + \lambda_5 P_{S5}(t) + \lambda_6 P_{S6}(t) + \lambda_7 P_{S7}(t) + \lambda_8 P_{S8}(t)]
\end{aligned}
\tag{5.165}
$$

we obtain differential equation of the first order:

$$\frac{dP_{S0}(t)}{dt} = -(\lambda_1 + \lambda_2 + \lambda_3 + \lambda_4 + \lambda_5 + \lambda_6 + \lambda_7 + \lambda_8)P_{S0}(t)$$
$$+ \lambda_1 P_{S1}(t) + \lambda_2 P_{S2}(t) + \lambda_3 P_{S3}(t) + \lambda_4 P_{S4}(t) + \lambda_5 P_{S5}(t) \qquad (5.166)$$
$$+ \lambda_6 P_{S6}(t) + \lambda_7 P_{S7}(t) + \lambda_8 P_{S8}(t)$$

Similar derivations are in the appendix. With the derivation, we obtain the following differential equations. For S_1:

$$\frac{dP_{S1}(t)}{dt} = \mu_1 P_{S0}(t) - (\lambda_1 + \mu_1)P_{S1}(t) + \lambda_1 P_{SF}(t) \qquad (5.167)$$

Equation for S_2 is:

$$\frac{dP_{S2}(t)}{dt} = \mu_2 P_{S0}(t) - (\lambda_2 + \mu_2)P_{S2}(t) + \lambda_2 P_{SF}(t) \qquad (5.168)$$

Equation for S_3 is:

$$\frac{dP_{S3}(t)}{dt} = \mu_3 P_{S0}(t) - (\lambda_3 + \alpha_{83} + \mu_3)P_{S3}(t) + \alpha_{83} P_{S8}(t) + \lambda_3 P_{SF}(t) \qquad (5.169)$$

Differential equation for S_4 is:

$$\frac{dP_{S4}(t)}{dt} = \mu_4 P_{S0}(t) - (\lambda_4 + \mu_4 + \alpha_{54} + \alpha_{74} + \alpha_{84})P_{S4}(t)$$
$$+ \alpha_{54} P_{S5}(t) + \alpha_{74} P_{S7}(t) + \alpha_{84} P_{S8}(t) + \lambda_4 P_{SF}(t) \qquad (5.170)$$

Differential equation for S_5 is:

$$\frac{dP_{S5}(t)}{dt} = \mu_5 P_{S0}(t) + \alpha_{15} P_{S1}(t) - (\lambda_5 + \mu_5 + \alpha_{15} + \alpha_{81})P_{S5}(t)$$
$$+ \alpha_{85} P_{S8}(t) + (\mu_5 + \alpha_{54})P_{SF}(t) \qquad (5.171)$$

Differential equation for S_6 is:

$$\frac{dP_{S6}(t)}{dt} = \mu_6 P_{S0}(t) + \alpha_{26} P_{S2}(t)$$
$$- (\lambda_6 + \alpha_{26} + \alpha_{86} + \mu_6) P_{S6}(t) + \alpha_{86} P_{S8}(t) + \mu_6 P_{SF} \tag{5.172}$$

Differential equation for S_7 is:

$$\frac{dP_{S7}(t)}{dt} = \mu_7 P_{S0}(t) - (\mu_7 + \lambda_7 + \alpha_{87}) P_{S7}(t)$$
$$+ (\mu_8 + \alpha_{83} + \alpha_{84} + \alpha_{85} + \alpha_{86} + \alpha_{87}) P_{S8}(t) + \lambda_7 P_{SF}(t) \tag{5.173}$$

Differential equation for S_8 is:

$$\frac{dP_{S8}(t)}{dt} = \mu_8 P_{S0}(t) - (\mu_8 + \lambda_8) P_{S8}(t) + \lambda_8 P_{SF}(t) \tag{5.174}$$

Differential equation for S_F is the most complex and is equal to:

$$\frac{dP_{SF}(t)}{dt} = (\mu_1 + \alpha_{15}) P_{S1}(t) + (\mu_2 + \alpha_{26}) P_{S2}(t) + \mu_3 P_{S3}(t)$$
$$+ \mu_4 P_{S4}(t) + (\mu_5 + \alpha_{54}) P_{S5}(t) + \mu_6 P_{S6}(t)$$
$$+ (\mu_7 + \alpha_{74}) P_{S7}(t) + (\mu_8 + \alpha_{83} + \alpha_{84} + \alpha_{85} + \alpha_{86} + \alpha_{87}) P_{S8}(t)$$
$$- (\mu_1 + \alpha_{15} + \mu_2 + \alpha_{26} + \mu_3 + \mu_4 + \mu_5 + \alpha_{54} + \mu_6$$
$$+ \mu_7 + \alpha_{74} + \mu_8 + \alpha_{83} + \alpha_{84} + \alpha_{85} + \alpha_{86} + \alpha_{87}) P_{SF}(t) \tag{5.175}$$

Since, the above differential equations are interconnected, they represent a system of 10 differential equations of the first order. This can be rewritten to matrix equation:

$$
\begin{bmatrix}
\frac{dP_{S0}}{dt} \\[2pt]
\frac{dP_{S1}}{dt} \\[2pt]
\frac{dP_{S2}}{dt} \\[2pt]
\frac{dP_{S3}}{dt} \\[2pt]
\frac{dP_{S4}}{dt} \\[2pt]
\frac{dP_{S5}}{dt} \\[2pt]
\frac{dP_{S6}}{dt} \\[2pt]
\frac{dP_{S7}}{dt} \\[2pt]
\frac{dP_{S8}}{dt} \\[2pt]
\frac{dP_{SF}}{dt}
\end{bmatrix}
=
$$

$$
\begin{bmatrix}
-\!\left(\begin{smallmatrix}\lambda_1+\lambda_2+\lambda_3\\+\lambda_4+\lambda_5+\lambda_6\\+\lambda_7+\lambda_8\end{smallmatrix}\right) & \lambda_1 & \lambda_2 & \lambda_3 & \lambda_4 & \lambda_5 & \lambda_6 & \lambda_7 & \lambda_8 & 0 \\[6pt]
\mu_1 & -(\lambda_1+\mu_1) & 0 & 0 & 0 & 0 & 0 & 0 & 0 & \lambda_1 \\[4pt]
\mu_2 & 0 & -(\lambda_2+\mu_2) & 0 & 0 & 0 & 0 & 0 & 0 & \lambda_2 \\[4pt]
\mu_3 & 0 & 0 & -\!\left(\begin{smallmatrix}\lambda_3+\alpha_{83}\\+\mu_3\end{smallmatrix}\right) & 0 & 0 & 0 & 0 & \alpha_{83} & \lambda_3 \\[6pt]
\mu_4 & 0 & 0 & 0 & -\!\left(\begin{smallmatrix}\lambda_4+\mu_4\\+\alpha_{54}+\alpha_{74}\\+\alpha_{84}\end{smallmatrix}\right) & \alpha_{54} & 0 & \alpha_{74} & \alpha_{84} & \lambda_4 \\[8pt]
\mu_5 & \alpha_{15} & 0 & 0 & 0 & -\!\left(\begin{smallmatrix}\lambda_5+\mu_5\\+\alpha_{15}+\alpha_{81}\end{smallmatrix}\right) & 0 & 0 & \alpha_{85} & (\mu_5+\alpha_{54}) \\[6pt]
\mu_6 & 0 & \alpha_{26} & 0 & 0 & 0 & -\!\left(\begin{smallmatrix}\lambda_6+\alpha_{26}\\+\alpha_{86}+\mu_6\end{smallmatrix}\right) & 0 & \alpha_{86} & \mu_6 \\[6pt]
\mu_7 & 0 & 0 & 0 & 0 & 0 & 0 & -(\mu_7+\lambda_7+\alpha_{87}) & \left(\begin{smallmatrix}\mu_8+\alpha_{83}\\+\alpha_{84}+\alpha_{85}\\+\alpha_{86}+\alpha_{87}\end{smallmatrix}\right) & \lambda_7 \\[8pt]
\mu_8 & 0 & 0 & 0 & 0 & 0 & 0 & 0 & -(\mu_8+\lambda_8) & \lambda_8 \\[4pt]
0 & (\mu_1+\alpha_{15}) & (\mu_2+\alpha_{26}) & \mu_3 & \mu_4 & (\mu_5+\alpha_{54}) & \mu_6 & (\mu_7+\alpha_{74}) & \left(\begin{smallmatrix}\mu_8+\alpha_{83}\\+\alpha_{84}+\alpha_{85}\\+\alpha_{86}+\alpha_{87}\end{smallmatrix}\right) & \left(\begin{smallmatrix}\mu_1+\alpha_{15}+\mu_2\\+\alpha_{26}+\mu_3+\mu_4\\+\mu_5+\alpha_{54}+\mu_6\\+\mu_7+\alpha_{74}\\+\mu_8+\alpha_{83}+\alpha_{84}\\+\alpha_{85}+\alpha_{86}+\alpha_{87}\end{smallmatrix}\right)
\end{bmatrix}
\begin{bmatrix}
P_{S0} \\
P_{S1} \\
P_{S2} \\
P_{S3} \\
P_{S4} \\
P_{S5} \\
P_{S6} \\
P_{S7} \\
P_{S8} \\
P_{SF}
\end{bmatrix}
\tag{5.176}
$$

The matrix equation (5.176) is used for Matlab/Simulink solution.

Chapter 6
Conclusions

Abstract Conclusions of the presented research were mentioned in this chapter: a framework, a noise influence, solutions for the reliability and the availability of the system, and the observed tool's limitation in solving the availability equations.

Keywords Software tool limitation · Framework · Assessing reliability · Wavelet image quality control system

This work presents a framework for assessing reliability and availability of visual quality control systems with example in wavelet computer vision. Special attention is given to influencing parameters, which can severely impact the results of quality systems and wavelet computer vision algorithm.

Performance of wavelet computer vision algorithm is explored in Chap. 3. Table 3.1 shows original experimental results for the used algorithm. It is analyzed how Gaussian, Salt & Pepper and Speckle noise influence the algorithm performance and tolerance to the error. It is shown that the highest tolerance of the algorithm is to speckle noise (25.167 %) and the lowest to the Salt & Pepper noise.

Hardware and software influencing parameters were discussed in Chap. 4. Analysis of relationships between influencing parameters was taken into account when developing model for reliability and model for availability.

Analytical results of the system of differential equations are obtained in Matlab for reliability case. There are generalized and particular solutions can be obtained by substituting real numbers from the manufacturer of interest.

Every research opens questions and this one is no exception. It is disputable whether the usage of 0 and 1 for state values is applicable in this case. This approach assumes that a given parameter can have a conclusively good or bad influence on system's performances. But the reality is not always black and white, but rather gray. This leads to possible guidelines and ideas for further work. States could be modeled as fuzzy states. Alternatively, several grades could be implemented for every state. This approach would make it harder to solve the problem, because the application of the true table approach would not be possible.

© The Author(s) 2015
I. Kuzmanić and I. Vujović, *Reliability and Availability of Quality Control Based on Wavelet Computer Vision*, SpringerBriefs in Electrical and Computer Engineering, DOI 10.1007/978-3-319-13317-1_6

Unfortunately, we could not solve the availability differential equations with our Matlab resources ("out of memory error") and find direct solutions for availabilities. However, we succeeded in developing a Simulink numerical solution.

The problem of resolution of the reliability problem was easily dealt with by Matlab Symbolic Toolbox and we found direct solutions for all reliabilities.

Appendix

A.1 Solving Differential Equations for Reliability in Matlab

The solutions to equations are obtained in Matlab by the following commands. The probability of state 7, P_{S7} is calculated by:

```
>> syms p7 lambda7 alpha87 lambda8 dt % definition of symbolic variables
>> dsolve('Dp7 + d*p7 = lambda7 + alpha87*lambda8*t*exp(–lambda8*t)')
% solving differential equation
```

The computer response was:

```
ans =
C2/exp(d*t) + ((lambda7*exp(d*t))/d – (alpha87*lambda8*exp(d*t – lambda8*t))/(d –
lambda8)^2 + (alpha87*lambda8*t*exp(d*t – lambda8*t))/(d – lambda8))/exp(d*t)
```

This command line answer can be written as:

$$
\begin{aligned}
P_{S7} = C_2 e^{-dt} + \left(\left(\lambda_7 \cdot e^{dt} \right)/d - \left(\alpha_{87} \cdot \lambda_8 e^{(dt - \lambda_8 t)} \right) \right)/(d - \lambda_8)^2 \\
+ \left(\alpha_{87} \cdot \lambda_8 t e^{(dt - \lambda_8 t)} \right)/(d - \lambda_8) e^{-dt}
\end{aligned}
\tag{A.1}
$$

$$
\begin{aligned}
P_{S7} = C_2 e^{-(\lambda_7 + \alpha_{87})t} \\
+ \left(\left(\lambda_7 \cdot e^{(\lambda_7 + \alpha_{87})t} \right)/(\lambda_7 + \alpha_{87}) - \left(\alpha_{87} \cdot \lambda_8 e^{((\lambda_7 + \alpha_{87})t - \lambda_8 t)} \right) \right) \\
/((\lambda_7 + \alpha_{87}) - \lambda_8)^2 \\
+ \left(\alpha_{87} \cdot \lambda_8 t e^{((\lambda_7 + \alpha_{87})t - \lambda_8 t)} \right)/((\lambda_7 + \alpha_{87}) - \lambda_8) e^{-(\lambda_7 + \alpha_{87})t}
\end{aligned}
\tag{A.2}
$$

$$
\begin{aligned}
P_{S7} = C_2 e^{-(\lambda_7 + \alpha_{87})t} \\
+ \frac{\left(\lambda_7 \cdot e^{(\lambda_7 + \alpha_{87})t} \right)/(\lambda_7 + \alpha_{87}) - \left(\alpha_{87} \cdot \lambda_8 e^{((\lambda_7 + \alpha_{87})t - \lambda_8 t)} \right)}{(\lambda_7 + \alpha_{87} - \lambda_8)^2} \\
+ \frac{\alpha_{87} \cdot \lambda_8 t e^{((\lambda_7 + \alpha_{87})t - \lambda_8 t)}}{(\lambda_7 + \alpha_{87}) - \lambda_8} e^{-(\lambda_7 + \alpha_{87})t}
\end{aligned}
\tag{A.3}
$$

© The Author(s) 2015
I. Kuzmanić and I. Vujović, *Reliability and Availability of Quality Control Based on Wavelet Computer Vision*, SpringerBriefs in Electrical and Computer Engineering, DOI 10.1007/978-3-319-13317-1

$$P_{S7} = C_2 e^{-(\lambda_7 + \alpha_{87})t} + \frac{\frac{\lambda_7 \cdot e^{(\lambda_7 + \alpha_{87})t}}{\lambda_7 + \alpha_{87}} - \alpha_{87} \cdot \lambda_8 e^{(\lambda_7 - \lambda_8 + \alpha_{87})t}}{(\lambda_7 + \alpha_{87} - \lambda_8)^2}$$

$$+ \frac{\alpha_{87} \cdot \lambda_8 t e^{(\lambda_7 - \lambda_8 + \alpha_{87})t}}{(\lambda_7 + \alpha_{87}) - \lambda_8} e^{-(\lambda_7 + \alpha_{87})t}$$

$$= C_2 e^{-(\lambda_7 + \alpha_{87})t} + \frac{\frac{\lambda_7 \cdot e^{(\lambda_7 + \alpha_{87})t}}{\lambda_7 + \alpha_{87}} - \alpha_{87} \cdot \lambda_8 e^{(\lambda_7 - \lambda_8 + \alpha_{87})t}}{(\lambda_7 + \alpha_{87} - \lambda_8)^2}$$

$$+ \frac{\alpha_{87} \cdot \lambda_8 t e^{-\lambda_8 t}}{\lambda_7 + \alpha_{87} - \lambda_8} \tag{A.4}$$

where C_2 is integration constant. In order to calculate exact solution, initial condition, based on Table 5.2, is used. According to Table 5.2, initial value of P_{S7} is 0.

```
>>dsolve('Dp7 + d*p7 = lambda7 + alpha87*lambda8*t*exp(–lambda8*t)',' p7(0) = 0')
ans =
((lambda7*exp(d*t))/d – (alpha87*lambda8*exp(d*t – lambda8*t))/(d – lambda8)^2
+ (alpha87*lambda8*t*exp(d*t – lambda8*t))/(d – lambda8))/exp(d*t) – (lambda7/d
– (alpha87*lambda8)/(d – lambda8)^2)/exp(d*t)
```

which finally give us the expression:

$$P_{S7} = \left(\left(\lambda_7 e^{dt} \right)/d - \left(\alpha_{87} \lambda_8 e^{dt - \lambda_8 t} \right)/(d - \lambda_8)^2 \right.$$
$$\left. + (\alpha_{87} \lambda_8 t e^{(dt - \lambda_8 t +)})/(d - \lambda_8) \right)/e^{dt} \tag{A.5}$$
$$- \left(\lambda_7/d - (\alpha_{87} \lambda_8)/(d - \lambda_8)^2 \right) e^{dt}$$

Adjusting to easy-view, we obtain:

$$P_{S7} = \left(\frac{\lambda_7 e^{(\lambda_7 + \alpha_{87})t}}{\lambda_7 + \alpha_{87}} - \frac{\alpha_{87} \lambda_8 e^{(\lambda_7 + \alpha_{87} - \lambda_8)t}}{(\lambda_7 + \alpha_{87} - \lambda_8)^2} + \frac{\alpha_{87} \lambda_8 t e^{((\lambda_7 + \alpha_{87} - \lambda_8)t)}}{\lambda_7 + \alpha_{87} - \lambda_8} \right)$$
$$\cdot e^{-(\lambda_7 + \alpha_{87})t} \tag{A.6}$$
$$- \left(\frac{\lambda_7}{\lambda_7 + \alpha_{87}} - \frac{\alpha_{87} \lambda_8}{(\lambda_7 + \alpha_{87} - \lambda_8)^2} \right) e^{(\lambda_7 + \alpha_{87})t}$$

Since the equation for probability P_{S6} is of a different type, we have to express it in another form. First, we must express limit and then differential:

$$\lim_{\Delta t \to 0} \frac{P_{S6}(t + \Delta t) - P_{S6}(t)}{\Delta t} = \lim_{\Delta t \to 0} [\lambda_6 P_{S0}(t) + [-(\lambda_6 + \alpha_{26} + \alpha_{86})] P_{S6}(t)$$
$$+ \alpha_{26} P_{S2}(t) + \alpha_{86} P_{S8}(t)] \tag{A.7}$$

$$\frac{dP_{S6}(t)}{dt} = \lambda_6 P_{S0}(t) - (\lambda_6 + \alpha_{26} + \alpha_{86})P_{S6}(t) + \alpha_{26}P_{S2}(t) + \alpha_{86}P_{S8}(t) \qquad \text{(A.8)}$$

$$\frac{dP_{S6}(t)}{dt} + (\lambda_6 + \alpha_{26} + \alpha_{86})P_{S6}(t) = \lambda_6 + \alpha_{26}P_{S2}(t) + \alpha_{86}P_{S8}(t) \qquad \text{(A.9)}$$

It should be noted that the solution for P_{S6} can only be obtained if solutions for P_{S2} and P_{S8} are entered. It is assumed that the initial value of P_{S6} is equal to zero.

```
>> syms p6 lambda6 alpha26 alpha86 lambda2 t lambda8
>> dsolve('Dp6 + (lambda6 + alpha26 + alpha86)*p6 = lambda6 - apha26 *lamb-
da2*t*exp(-lambda2*t) + alpha86*lambda8*t*exp(-lambda8*t)',' p6(0) = 0')
```

The Matlab response was:

ans =
((lambda6*exp(alpha26*t + alpha86*t + lambda6*t))/(alpha26 + alpha86 + lambda6)
+ (alpha26*lambda2*exp(alpha26*t + alpha86*t – lambda2*t + lambda6*t))/
(alpha26 + alpha86 – lambda2 + lambda6)^2 – (alpha86*lambda8*exp(alpha26*t +
alpha86*t + lambda6*t – lambda8*t))/(alpha26 + alpha86 + lambda6 – lambda8)^2 –
(alpha26*lambda2*t*exp(alpha26*t + alpha86*t – lambda2*t + lambda6*t))/
(alpha26 + alpha86 – lambda2 + lambda6) + (alpha86*lambda8*t*exp(alpha26*t +
alpha86*t + lambda6*t – lambda8*t))/(alpha26 + alpha86 + lambda6 – lambda8))/
exp(t*(alpha26 + alpha86 + lambda6)) – (lambda6/(alpha26 + alpha86 + lambda6) +
(alpha26*lambda2)/(alpha26 + alpha86 – lambda2 + lambda6)^2 – (alpha86*-
lambda8)/(alpha26 + alpha86 + lambda6 – lambda8)^2)/exp(t*(alpha26 + alpha86 +
lambda6))

which can be written prettier as in the following equation:

$$
\begin{aligned}
P_{S6} = &\left(\frac{\lambda_6 e^{\alpha_{26}t+\alpha_{86}t+\lambda_6 t}}{\alpha_{26} + \alpha_{86} + \lambda_6} + \frac{\alpha_{26}\lambda_2 e^{\alpha_{26}t+\alpha_{86}t-\lambda_2 t+\lambda_6 t}}{(\alpha_{26} + \alpha_{86} - \lambda_2 + \lambda_6)^2} \right. \\
&- \frac{\alpha_{86}\lambda_8 e^{(\alpha_{26}+\alpha_{86}+\lambda_6-\lambda_8)t}}{(\alpha_{26} + \alpha_{86} + \lambda_6 - \lambda_8)^2} - \frac{\alpha_{26}\lambda_2 t e^{\alpha_{26}t+\alpha_{86}t-\lambda_2 t+\lambda_6 t}}{\alpha_{26} + \alpha_{86} - \lambda_2 + \lambda_6} \\
&\left. + \frac{\alpha_{86}\lambda_8 t e^{\alpha_{26}t+\alpha_{86}t+\lambda_6 t-\lambda_8 t}}{\alpha_{26} + \alpha_{86} + \lambda_6 - \lambda_8} \right) e^{-t(\alpha_{26}+\alpha_{86}-\lambda_6)} \\
&- \left(\frac{\lambda_6}{\alpha_{26} + \alpha_{86} + \lambda_6} + \frac{\alpha_{26}\lambda_2}{(\alpha_{26} + \alpha_{86} + \lambda_6 - \lambda_2)^2} \right. \\
&\left. - \frac{\alpha_{86}\lambda_8}{(\alpha_{26} + \alpha_{86} + \lambda_6 - \lambda_8)^2} \right) \cdot e^{-t(\alpha_{26}+\alpha_{86}+\lambda_6)}
\end{aligned} \qquad \text{(A.10)}
$$

The same procedure was followed to obtain the solution for P_{S5}.

$$\lim_{\Delta t \to 0} \frac{P_{S5}(t + \Delta t) - P_{S5}(t)}{\Delta t} = \lim_{\Delta t \to 0} [\lambda_5 P_{S0}(t)$$
$$+ [-(\lambda_5 + \alpha_{15} + \alpha_{81})]P_{S5}(t) + \alpha_{14}P_{S1}(t) + \alpha_{85}P_{S8}(t)] \tag{A.11}$$

$$\frac{dP_{S5}(t)}{dt} = \lambda_5 P_{S0}(t) - (\lambda_5 + \alpha_{15} + \alpha_{81})P_{S5}(t)$$
$$+ \alpha_{14}P_{S1}(t) + \alpha_{85}P_{S8}(t) \tag{A.12}$$

$$\frac{dP_{S5}(t)}{dt} + (\lambda_5 + \alpha_{15} + \alpha_{81})P_{S5}(t) = \lambda_5 + \alpha_{14}P_{S1}(t) + \alpha_{85}P_{S8}(t) \tag{A.13}$$

which means that three equations have 3 unknowns. All chances are that 2 of them have already been calculated.

For the Matlab solution, the initial value of P_{S5} is also assumed to be equal to zero.

```
>> syms p5 lambda2 alpha15 alpha81 lambda5 alpha14 lambda1 t alpha85
lambda8
>> dsolve('Dp5 + (lambda2 + alpha15 + alpha81)*p5 = lambda5 + alpha14*-
lambda1*t*exp(−lambda1*t) + alpha85*lambda8*t*exp(−lambda8*t)', 'p5(0) = 0')
```

The Matlab response was:

```
ans =
((alpha14*lambda1)/(alpha15 + alpha81 − lambda1 + lambda2)^2 − lambda5/
(alpha15 + alpha81 + lambda2) + (alpha85*lambda8)/(alpha15 + alpha81 + lambda2
− lambda8)^2)/exp(t*(alpha15 + alpha81 + lambda2)) + ((lambda5*exp(alpha15*t +
alpha81*t + lambda2*t))/(alpha15 + alpha81 + lambda2) − (alpha14*lambda1*exp
(alpha15*t + alpha81*t − lambda1*t + lambda2*t))/(alpha15 + alpha81 − lambda1 +
lambda2)^2 − (alpha85*lambda8*exp(alpha15*t + alpha81*t + lambda2*t − lamb-
da8*t))/(alpha15 + alpha81 + lambda2 − lambda8)^2 + (alpha14*lambda1*t*exp
(alpha15*t + alpha81*t − lambda1*t + lambda2*t))/(alpha15 + alpha81 − lambda1 +
lambda2) + (alpha85*lambda8*t*exp(alpha15*t + alpha81*t + lambda2*t − lamb-
da8*t))/(alpha15 + alpha81 + lambda2 − lambda8))/exp(t*(alpha15 + alpha81 +
lambda2))
```

which can be written as:

$$P_{S5} = \left(\frac{\alpha_{14}\lambda_1}{(\alpha_{15} + \alpha_{81} - \lambda_1 + \lambda_2)^2} - \frac{\lambda_5}{\alpha_{15} + \alpha_{81} + \lambda_2} \right.$$

$$\left. + (\alpha_{85}\lambda_8)/(\alpha_{15} + \alpha_{81} + \lambda_2 - \lambda_8)^2 \right)/e^{t(\alpha_{15}+\alpha_{81}+\lambda_2)}$$

$$+ \left(\frac{\lambda_5 e^{(\alpha_{15}+\alpha_{81}+\lambda_2)t}}{\alpha_{15} + \alpha_{81} + \lambda_2} - \left(\alpha_{14}\lambda_1 e^{(\alpha_{15}+\alpha_{81}-\lambda_1+\lambda_2)t} \right) \right.$$

$$/(\alpha_{15} + \alpha_{81} - \lambda_1 + \lambda_2)^2 - \left(\alpha_{85}\lambda_8 e^{(\alpha_{15}+\alpha_{81}+\lambda_2-\lambda_8)t} \right) \qquad (A.14)$$

$$/(\alpha_{15} + \alpha_{81} + \lambda_2 - \lambda_8)^2 + \left(\alpha_{14}\lambda_1 t e^{(\alpha_{15}+\alpha_{81}-\lambda_1+\lambda_2)t} \right)$$

$$/(\alpha_{15} + \alpha_{81} - \lambda_1 + \lambda_2)$$

$$\left. + \frac{\alpha_{85}\lambda_8 t e^{(\alpha_{15}+\alpha_{81}+\lambda_2-\lambda_8)t}}{\alpha_{15}\alpha_{81} + \lambda_2 - \lambda_8} \right)/e^{t(\alpha_{15}+\alpha_{81}+\lambda_2)}$$

or by further rearranging:

$$P_{S5} = \left(\frac{\alpha_{14}\lambda_1}{(\alpha_{15} + \alpha_{81} - \lambda_1 + \lambda_2)^2} - \frac{\lambda_5}{\alpha_{15} + \alpha_{81} + \lambda_2} \right.$$

$$\left. + \frac{\alpha_{85}\lambda_8}{(\alpha_{15} + \alpha_{81} + \lambda_2 - \lambda_8)^2} \right) \cdot e^{-t(\alpha_{15}+\alpha_{81}+\lambda_2)}$$

$$+ \left(\frac{\lambda_5 e^{\alpha_{15}t+\alpha_{81}t+\lambda_2 t}}{\alpha_{15} + \alpha_{81} + \lambda_2} - \frac{\alpha_{14}\lambda_1 e^{\alpha_{15}t+\alpha_{81}t-\lambda_1 t+\lambda_2 t}}{(\alpha_{15} + \alpha_{81} - \lambda_1 + \lambda_2)^2} \right.$$

$$- \frac{\alpha_{85}\lambda_8 e^{\alpha_{15}t+\alpha_{81}t+\lambda_2 t-\lambda_8 t}}{(\alpha_{15} + \alpha_{81} + \lambda_2 - \lambda_8)^2}$$

$$\left. + \frac{\alpha_{14}\lambda_1 t e^{\alpha_{15}t+\alpha_{81}t-\lambda_1 t+\lambda_2 t}}{\alpha_{15} + \alpha_{81} - \lambda_1 + \lambda_2} + \frac{\alpha_{85}\lambda_8 t e^{\alpha_{15}t+\alpha_{81}t+\lambda_2 t-\lambda_8 t}}{\alpha_{15} + \alpha_{81} + \lambda_2 - \lambda_8} \right) e^{-t(\alpha_{15}+\alpha_{81}+\lambda_2)}$$

$$(A.15)$$

Further, we can derive the differential equation for P_{S4}:

$$\lim_{\Delta t \to 0} \frac{P_{S4}(t + \Delta t) - P_{S4}(t)}{\Delta t} = \lim_{\Delta t \to 0} [\lambda_4 P_{S0}(t) + [-(\lambda_4 + \alpha_{54} + \alpha_{74} + \alpha_{84})]P_{S4}(t)$$

$$+ \alpha_{54}P_{S5}(t) + \alpha_{74}P_{S7}(t) + \alpha_{84}P_{S8}(t)]$$

$$(A.16)$$

The results are the differential equation of the first order:

$$\frac{dP_{S4}(t)}{dt} = \lambda_4 P_{S0}(t) - (\lambda_4 + \alpha_{54} + \alpha_{74} + \alpha_{84})P_{S4}(t)$$
$$+ \alpha_{54}P_{S5}(t) + \alpha_{74}P_{S7}(t) + \alpha_{84}P_{S8}(t) \tag{A.17}$$

which can be written in known form as:

$$\frac{dP_{S4}(t)}{dt} + (\lambda_4 + \alpha_{54} + \alpha_{74} + \alpha_{84})P_{S4}(t) = \lambda_4 + \alpha_{54}P_{S5}(t) + \alpha_{74}P_{S7}(t) + \alpha_{84}P_{S8}(t)$$
$$\tag{A.18}$$

In order to calculate P_{S4}, the solution for P_{S5} must be present in the workspace. Otherwise, it should be entered before the next "DSolve" line. P_{S4} is calculated in Matlab by the following lines:

```
>> syms p4 lambda4 alpha54 alpha74 alpha84 lambda8 t
>> dsolve ('Dp4 + (lambda4 + alpha54 + alpha74 + alpha84)*p4 = lambda4 +
alpha54*p5 + alpha74*lambda8*t*exp(-lambda8*t)', 'p4(0) = 0')
```

The Matlab response was:

```
ans =
((exp(al54*t + al74*t + al84*t + la4*t)*(la4 + al54*ans))/(al54 + al74 + al84 + la4)
- (al74*la8*exp(al54*t + al74*t + al84*t + la4*t - la8*t))/(al54 + al74 + al84 + la4
- la8)^2 + (al74*la8*t*exp(al54*t + al74*t + al84*t + la4*t - la8*t))/(al54 + al74 +
al84 + la4 - la8))/exp(t*(al54 + al74 + al84 + la4)) - ((la4 + al54*ans)/(al54 + al74
+ al84 + la4) - (al74*la8)/(al54 + al74 + al84 + la4 - la8)^2)/exp(t*(al54 + al74 +
al84 + la4))
```

which can be expressed as:

$$P_{S4} = \Big(\big(e^{\alpha_{54}t+\alpha_{74}t+\alpha_{84}t+\lambda_4 t}\big)\big(\lambda_4 + \alpha_{54}\cdot P_{S5}\big)\Big)/(\alpha_{54} + \alpha_{74} + \alpha_{84} + \lambda_4)$$
$$- \big(\alpha_{74}\lambda_8 e^{\alpha_{54}t+\alpha_{74}t+\alpha_{84}t+\lambda_4 t-\lambda_8 t}\big)$$
$$/(\alpha_{54} + \alpha_{74} + \alpha_{84} + \lambda_4 - \lambda_8)^2 + \big(\alpha_{74}\lambda_8 t e^{\alpha_{54}t+\alpha_{74}t+\alpha_{84}t+\lambda_4 t-\lambda_8 t}\big)$$
$$/(\alpha_{54} + \alpha_{74} + \alpha_{84} + \lambda_4 - \lambda_8)/e^{t(\alpha_{54}+\alpha_{74}+\alpha_{84}+\lambda_4)} \tag{A.19}$$
$$- \big((\lambda_4 + \alpha_{54}P_{S5})/(\alpha_{54} + \alpha_{74} + \alpha_{84} + \lambda_4)$$
$$-(\alpha_{74}\lambda_8)/(\alpha_{54} + \alpha_{74} + \alpha_{84} + \lambda_4 - \lambda_8)^2\big)/e^{(\alpha_{54}+\alpha_{74}+\alpha_{84}+\lambda_4)t}$$

or further rearranging:

$$
\begin{aligned}
P_{S4} = \Bigg(& \frac{(\lambda_4 + \alpha_{54} \cdot P_{S5})e^{(\alpha_{54}+\alpha_{74}+\alpha_{84}+\lambda_4)t}}{\alpha_{54} + \alpha_{74} + \alpha_{84} + \lambda_4} \Bigg) \\
& - \frac{\alpha_{74}\lambda_8 e^{(\alpha_{54}+\alpha_{74}+\alpha_{84}+\lambda_4-\lambda_8)t}}{(\alpha_{54} + \alpha_{74} + \alpha_{84} + \lambda_4 - \lambda_8)^2} \\
& + \frac{\alpha_{74}\lambda_8 t e^{(\alpha_{54}+\alpha_{74}+\alpha_{84}+\lambda_4-\lambda_8)t}}{\alpha_{54} + \alpha_{74} + \alpha_{84} + \lambda_4 - \lambda_8} e^{-t(\alpha_{54}+\alpha_{74}+\alpha_{84}+\lambda_4)} \\
& - \Bigg(\frac{\lambda_4 + \alpha_{54}P_{S5}}{\alpha_{54} + \alpha_{74} + \alpha_{84} + \lambda_4} \\
& - \frac{\alpha_{74}\lambda_8}{(\alpha_{54} + \alpha_{74} + \alpha_{84} + \lambda_4 - \lambda_8)^2} \Bigg) \cdot e^{-(\alpha_{54}+\alpha_{74}+\alpha_{84}+\lambda_4)t}
\end{aligned} \tag{A.20}
$$

P_{S4} can be derived in same manner:

$$
\lim_{\Delta t \to 0} \frac{P_{S3}(t + \Delta t) - P_{S3}(t)}{\Delta t} = \lim_{\Delta t \to 0} [\lambda_3 P_{S0}(t) + (-\lambda_3 - \alpha_{83})P_{S3}(t) + \alpha_{83}P_{S8}(t)] \tag{A.21}
$$

$$
\frac{dP_{S3}(t)}{dt} = \lambda_3 P_{S0}(t) - (\lambda_3 + \alpha_{83})P_{S3}(t) + \alpha_{83}P_{S8}(t) \tag{A.22}
$$

$$
\frac{dP_{S3}(t)}{dt} + (\lambda_3 + \alpha_{83})P_{S3}(t) = \lambda_3 \cdot 1 + \alpha_{83}P_{S8}(t) \tag{A.23}
$$

which is the equation with two unknowns. In order to solve it, we used lines:

```
>> syms p3 labmda3 alpha83 lambda8 t
>> dsolve ('Dp3 + (lambda3 + alpha83)*p3 = lambda3 + alpha83*lambda8*t*exp
(−lambda8*t)', 'p3(0) = 0')
```

where we assumed that initial condition for state 3, p(3) = 0. Matlab answer was:

```
ans =
((lambda3*exp(alpha83*t + lambda3*t))/(alpha83 + lambda3) − (alpha83 *
lambda8 * exp(alpha83*t + lambda3*t − lambda8*t))/(alpha83 + lambda3 −
lambda8)^2 + (alpha83*lambda8*t*exp(alpha83*t + lambda3*t − lambda8*t))/
(alpha83 + lambda3 − lambda8))/exp(t*(alpha83 + lambda3)) − (lambda3/(alpha83
+ lambda3) − (alpha83*lambda8)/(alpha83 + lambda3 − lambda8)^2)/exp(t*
(alpha83 + lambda3))
```

which can be written in human-perceptive manner as:

$$
\begin{aligned}
P_{S3} = & \left(\left(\lambda_3 e^{\alpha_{83}t + \lambda_3 t} \right) / (\alpha_{83} + \lambda_3) - \left(\alpha_{83}\lambda_8 e^{\alpha_{83}t + \lambda_3 t - \lambda_8 t} \right) \right. \\
& / (\alpha_{83} + \lambda_3 - \lambda_8)^2 \\
& \left. + \left(\alpha_{83}\lambda_8 t e^{\alpha_{83}t + \lambda_3 t - \lambda_8 t} \right) / (\alpha_{83} + \lambda_3 - \lambda_8) \right) / e^{t(\alpha_{83} + \lambda_3)} \\
& - \left(\lambda_3 / (\alpha_{83} + \lambda_3) - (\alpha_{83}\lambda_8) / (\alpha_{83} + \lambda_3 - \lambda_8)^2 \right) / e^{t(\alpha_{83} + \lambda_3)}
\end{aligned}
\tag{A.24}
$$

or further rearranging:

$$
\begin{aligned}
P_{S3} = & \left(\frac{\lambda_3 e^{(\alpha_{83} + \lambda_3)t}}{\alpha_{83} + \lambda_3} - \frac{\alpha_{83}\lambda_8 e^{(\alpha_{83} + \lambda_3 - \lambda_8)t}}{(\alpha_{83} + \lambda_3 - \lambda_8)^2} + \frac{\alpha_{83}\lambda_8 t e^{(\alpha_{83} + \lambda_3 - \lambda_8)t}}{\alpha_{83} + \lambda_3 - \lambda_8} \right) \\
& \cdot e^{-t(\alpha_{83} + \lambda_3)} \\
& - \left(\frac{\lambda_3}{\alpha_{83} + \lambda_3} - \frac{\alpha_{83}\lambda_8}{(\alpha_{83} + \lambda_3 - \lambda_8)^2} \right) e^{-t(\alpha_{83} + \lambda_3)}
\end{aligned}
\tag{A.25}
$$

The derivation procedure is similar, but with new expressions:

$$
\begin{aligned}
\lim_{\Delta t \to 0} \frac{P_{SF}(t + \Delta t) - P_{SF}(t)}{\Delta t} = \lim_{\Delta t \to 0} & [(\lambda_1 + \alpha_{15})P_{S1}(t) \\
& + (\lambda_2 + \alpha_{26})P_{S2}(t) + \lambda_3 P_{S3}(t) + \lambda_4 P_{S4}(t) + (\lambda_5 + \alpha_{54})P_{S5}(t) \\
& + \lambda_6 P_{S6}(t) + (\alpha_{74} + \lambda_7)P_{S7}(t) \\
& + (\lambda_8 + \alpha_{83} + \alpha_{84} + \alpha_{85} + \alpha_{86} + \alpha_{87})P_{S8}(t)]
\end{aligned}
\tag{A.26}
$$

and

$$
\begin{aligned}
\frac{dP_{SF}(t)}{dt} = & (\lambda_1 + \alpha_{15})P_{S1}(t) + (\lambda_2 + \alpha_{26})P_{S2}(t) \\
& + \lambda_3 P_{S3}(t) + \lambda_4 P_{S4}(t) \\
& + (\lambda_5 + \alpha_{54})P_{S5}(t) + \lambda_6 P_{S6}(t) + (\alpha_{74} + \lambda_7)P_{S7}(t) \\
& + (\lambda_8 + \alpha_{83} + \alpha_{84} + \alpha_{85} + \alpha_{86} + \alpha_{87})P_{S8}(t)
\end{aligned}
\tag{A.27}
$$

P_{SF} can be calculated if P_{S1}, P_{S2}, P_{S3}, P_{S4}, P_{S5}, P_{S6}, P_{S7} and P_{S8} are known from previous calculus, which means that the values must be in workspace, as well as all constants, α_{ij} and λ_j.

```
>> syms pf
>> dsolve ('Dpf = (lambda1 + alpha15)*p1 + (lambda2 + alpha26)*p2 + (lambda3)
*p3 + (lambda4)*p4 + (lambda5 + alpha54)*p5 + (lambda6)*p6 + (alpha74 +
lambda7)*p7 + (lambda3 + alpha83 + alpha84 + alpha85 + alpha86 + alpha87)*p8',
'pf(0) = 0')
```

The Matlab provided solution is:

ans =
t*(alpha15*p1 + alpha26*p2 + alpha54*p5 + alpha74*p7 + alpha83*p8 + alpha84*p8
+ alpha85*p8 + alpha86*p8 + alpha87*p8 + lambda1*p1 + lambda2*p2 + lambda3*p3
+ lambda4*p4 + lambda5*p5 + lambda6*p6 + lambda7*p7 + lambda3*p8)

that is:

$$
\begin{aligned}
P_{SF} = t \cdot (&\alpha_{15} \cdot P_{S1} + \alpha_{26} \cdot P_{S2} + \alpha_{54} \cdot P_{S5} + \alpha_{74} \cdot P_{S7} \\
&+ (\alpha_{83} + \alpha_{84} + \alpha_{85} + \alpha_{86} + \alpha_{87})P_{S8} + \lambda_1 P_{S1} \\
&+ \lambda_2 P_{S2} + \lambda_3 P_{S3} + \lambda_4 P_{S4} + \lambda_5 P_{S5} + \lambda_6 P_{S6} + \lambda_7 P_{S7} + \lambda_8 P_{S8})
\end{aligned}
\tag{A.28}
$$

A.2 Derivation of Differential Equations for Availability

The starting equation can be rewritten from Table 5.3:

$$
P_{S1}(t + \Delta t) = a_{01}P_{S0}(t) + a_{11}P_{S1}(t) + a_{F1}P_{SF}(t)
\tag{A.29}
$$

First, we should write a discrete-time version of the equation and then set the limit of Δt to zero:

$$
P_{S1}(t + \Delta t) = \mu_1 \Delta t P_{S0}(t) + [1 - \lambda_1 \Delta t - \mu_1 \Delta t]P_{S1}(t) + \lambda_1 \Delta t P_{SF}(t)
\tag{A.30}
$$

$$
\lim_{\Delta t \to 0} \frac{P_{S1}(t + \Delta t) - P_{S1}(t)}{\Delta t} = \lim_{\Delta t \to 0}[\mu_1 P_{S0}(t) - (\lambda_1 + \mu_1)P_{S1}(t) + \lambda_1 P_{SF}(t)]
\tag{A.31}
$$

Finally, we have:

$$
\frac{dP_{S1}(t)}{dt} = \mu_1 P_{S0}(t) - (\lambda_1 + \mu_1)P_{S1}(t) + \lambda_1 P_{SF}(t)
\tag{A.32}
$$

Similar procedure can be followed for S_2:

$$
P_{S2}(t + \Delta t) = a_{02}P_{S0}(t) + a_{22}P_{S2}(t) + a_{F2}P_{SF}(t)
\tag{A.33}
$$

Substituting coefficients:

$$
P_{S2}(t + \Delta t) = \mu_2 \Delta t P_{S0}(t) + [1 - \lambda_2 \Delta t - \mu_2 \Delta t]P_{S2}(t) + \lambda_2 \Delta t P_{SF}(t)
\tag{A.34}
$$

$$
\lim_{\Delta t \to 0} \frac{P_{S2}(t + \Delta t) - P_{S2}(t)}{\Delta t} = \lim_{\Delta t \to 0}[\mu_2 P_{S0}(t) - (\lambda_2 + \mu_2)P_{S2}(t) + \lambda_2 P_{SF}(t)]
\tag{A.35}
$$

$$\frac{dP_{S2}(t)}{dt} = \mu_2 P_{S0}(t) - (\lambda_2 + \mu_2)P_{S2}(t) + \lambda_2 P_{SF}(t) \qquad (A.36)$$

P_{S3} is derived by the following lines. First, we rewrite the discrete-time equation from Table 5.3:

$$P_{S3}(t + \Delta t) = a_{03}P_{S0}(t) + a_{33}P_{S3}(t) + a_{83}P_{S8}(t) + a_{F3}P_{SF}(t) \qquad (A.37)$$

Substituting for a_{03}, a_{33}, a_{83} and a_{F3}:

$$\begin{aligned} P_{S3}(t + \Delta t) = {} & \mu_3 \Delta t P_{S0}(t) + [1 - \lambda_3 \Delta t - \alpha_{83}\Delta t - \mu_3 \Delta t]P_{S3}(t) \\ & + \alpha_{83}\Delta t P_{S8}(t) + \lambda_3 \Delta t P_{SF}(t) \end{aligned} \qquad (A.38)$$

Once the limit is set, we get a differential equation in a continuous-time domain:

$$\lim_{\Delta t \to 0} \frac{P_{S3}(t + \Delta t) - P_{S3}(t)}{\Delta t} = \lim_{\Delta t \to 0}[\mu_3 P_{S0}(t) - (\lambda_3 + \alpha_{83} + \mu_3)P_{S3}(t) + \alpha_{83}P_{S8}(t) + \lambda_3 P_{SF}(t)]$$

$$(A.39)$$

$$\frac{dP_{S3}(t)}{dt} = \mu_3 P_{S0}(t) - (\lambda_3 + \alpha_{83} + \mu_3)P_{S3}(t) + \alpha_{83}P_{S8}(t) + \lambda_3 P_{SF}(t) \qquad (A.40)$$

P_{S4} is derived by firstly writing discrete-time equation:

$$\begin{aligned} P_{S4}(t + \Delta t) = {} & a_{04}P_{S0}(t) + a_{44}P_{S4}(t) + a_{54}P_{S5}(t) \\ & + a_{74}P_{S7}(t) + a_{84}P_{S8}(t) + a_{F4}P_{SF}(t) \end{aligned} \qquad (A.41)$$

And then substituting coefficients we obtain:

$$\begin{aligned} P_{S4}(t + \Delta t) = {} & \mu_4 \Delta t P_{S0}(t) \\ & + [1 - (\lambda_4 + \mu_4 + \alpha_{54} + \alpha_{74} + \alpha_{84})\Delta t]P_{S4}(t) \\ & + \alpha_{54}\Delta t P_{S5}(t) + \alpha_{74}\Delta t P_{S7}(t) + \alpha_{84}\Delta t P_{S8}(t) + \lambda_4 \Delta t P_{SF}(t) \end{aligned} \qquad (A.42)$$

Putting limit we get differential equation.

$$\begin{aligned} \lim_{\Delta t \to 0} \frac{P_{S4}(t + \Delta t) - P_{S4}(t)}{\Delta t} = {} & \lim_{\Delta t \to 0}[\mu_4 P_{S0}(t) - (\lambda_4 + \mu_4 + \alpha_{54} + \alpha_{74} + \alpha_{84})P_{S4}(t) \\ & + \alpha_{54}P_{S5}(t) + \alpha_{74}P_{S7}(t) + \alpha_{84}P_{S8}(t) + \lambda_4 P_{SF}(t)] \end{aligned}$$

$$(A.43)$$

$$\begin{aligned} \frac{dP_{S4}(t)}{dt} = {} & \mu_4 P_{S0}(t) - (\lambda_4 + \mu_4 + \alpha_{54} + \alpha_{74} + \alpha_{84})P_{S4}(t) \\ & + \alpha_{54}P_{S5}(t) + \alpha_{74}P_{S7}(t) + \alpha_{84}P_{S8}(t) + \lambda_4 P_{SF}(t) \end{aligned} \qquad (A.44)$$

P_{S5} is derived by the following lines:

$$P_{S5}(t + \Delta t) = a_{05}P_{S0}(t) + a_{15}P_{S1}(t) \\ + a_{55}P_{S5}(t) + a_{85}P_{S8}(t) + a_{F5}P_{SF}(t) \tag{A.45}$$

Substituting expressions for coefficients:

$$P_{S5}(t + \Delta t) = \mu_5 \Delta t P_{S0}(t) + \alpha_{15} \Delta t P_{S1}(t) \\ + [1 - (\lambda_5 + \mu_5 + \alpha_{15} + \alpha_{81}) \Delta t] P_{S5}(t) \tag{A.46} \\ + \alpha_{85} \Delta t P_{S8}(t) + [\mu_5 \Delta t + \alpha_{54} \Delta t] P_{SF}(t)$$

Putting limit:

$$\lim_{\Delta t \to 0} \frac{P_{S5}(t + \Delta t) - P_{S5}(t)}{\Delta t} = \lim_{\Delta t \to 0} [\mu_5 P_{S0}(t) + \alpha_{15} P_{S1}(t) \\ - (\lambda_5 + \mu_5 + \alpha_{15} + \alpha_{81}) P_{S5}(t) + \alpha_{85} P_{S8}(t) + (\mu_5 + \alpha_{54}) P_{SF}(t)] \tag{A.47}$$

Finally, we obtain:

$$\frac{dP_{S5}(t)}{dt} = \mu_5 P_{S0}(t) + \alpha_{15} P_{S1}(t) - (\lambda_5 + \mu_5 + \alpha_{15} + \alpha_{81}) P_{S5}(t) \\ + \alpha_{85} P_{S8}(t) + (\mu_5 + \alpha_{54}) P_{SF}(t) \tag{A.48}$$

P_{S6} is derived by the following lines:

$$P_{S6}(t + \Delta t) = a_{06}P_{S0}(t) + a_{26}P_{S2}(t) \\ + a_{66}P_{S6}(t) + a_{86}P_{S8}(t) + a_{F6}P_{SF}(t) \tag{A.49}$$

Substituting expressions for a_{06}, a_{26}, a_{66}, a_{86} and a_{F6}, we get:

$$P_{S6}(t + \Delta t) = \mu_6 \Delta t P_{S0}(t) + \alpha_{26} \Delta t P_{S2}(t) \\ + [1 - (\lambda_6 + \alpha_{26} + \alpha_{86} + \mu_6) \Delta t] P_{S6}(t) + \alpha_{86} \Delta t P_{S8}(t) + \mu_6 \Delta t P_{SF}(t) \tag{A.50}$$

Limit when time interval is infinitely small is:

$$\lim_{\Delta t \to 0} \frac{P_{S6}(t + \Delta t) - P_{S6}(t)}{\Delta t} = \lim_{\Delta t \to 0} [\mu_6 P_{S0}(t) + \alpha_{26} P_{S2}(t) \\ - (\lambda_6 + \alpha_{26} + \alpha_{86} + \mu_6) P_{S6}(t) + \alpha_{86} P_{S8}(t) + \mu_6 P_{SF}] \tag{A.51}$$

which is, in fact, differential equation in continuous time:

$$\frac{dP_{S6}(t)}{dt} = \mu_6 P_{S0}(t) + \alpha_{26} P_{S2}(t)$$
$$- (\lambda_6 + \alpha_{26} + \alpha_{86} + \mu_6) P_{S6}(t) + \alpha_{86} P_{S8}(t) + \mu_6 P_{SF} \tag{A.52}$$

P_{S7} is derived by the following lines:

$$P_{S7}(t + \Delta t) = a_{07} P_{S0}(t) + a_{77} P_{S7}(t) + a_{87} P_{S8}(t) + a_{F7} P_{SF}(t) \tag{A.53}$$

Substituting we obtain:

$$P_{S7}(t + \Delta t) = \mu_7 \Delta t P_{S0}(t) + [1 - (\lambda_7 + \mu_7 + \alpha_{87}) \Delta t] P_{S7}(t)$$
$$+ [\mu_8 \Delta t + (\alpha_{83} + \alpha_{84} + \alpha_{85} + \alpha_{86} + \alpha_{87}) \Delta t] P_{S8}(t) + \lambda_7 \Delta t P_{SF}(t) \tag{A.54}$$

Putting time interval to tend to zero and calculating limit, we have:

$$\lim_{\Delta t \to 0} \frac{P_{S7}(t + \Delta t) - P_{S7}(t)}{\Delta t} = \lim_{\Delta t \to 0} [\mu_7 P_{S0}(t) - (\mu_7 + \lambda_7 + \alpha_{87}) P_{S7}(t)$$
$$+ (\mu_8 + \alpha_{83} + \alpha_{84} + \alpha_{85} + \alpha_{86} + \alpha_{87}) P_{S8}(t) + \lambda_7 P_{SF}(t)] \tag{A.55}$$

and, finally:

$$\frac{dP_{S7}(t)}{dt} = \mu_7 P_{S0}(t) - (\mu_7 + \lambda_7 + \alpha_{87}) P_{S7}(t)$$
$$+ (\mu_8 + \alpha_{83} + \alpha_{84} + \alpha_{85} + \alpha_{86} + \alpha_{87}) P_{S8}(t) + \lambda_7 P_{SF}(t) \tag{A.56}$$

P_{S8} is derived by the following lines:

$$P_{S8}(t + \Delta t) = a_{08} P_{S0}(t) + a_{88} P_{S8}(t) + a_{F8} P_{SF}(t) \tag{A.57}$$

Substituting:

$$P_{S8}(t + \Delta t) = \mu_8 \Delta t P_{S0}(t) + [1 - (\mu_8 + \lambda_8) \Delta t] P_{S8}(t) + \lambda_8 \Delta t P_{SF}(t) \tag{A.58}$$

Putting limit:

$$\lim_{\Delta t \to 0} \frac{P_{S8}(t + \Delta t) - P_{S8}(t)}{\Delta t} = \lim_{\Delta t \to 0} [\mu_8 P_{S0}(t) - (\mu_8 + \lambda_8) P_{S8}(t) + \lambda_8 P_{SF}(t)] \tag{A.59}$$

Differential equation for S_8 is, therefore, expressed with:

$$\frac{dP_{S8}(t)}{dt} = \mu_8 P_{S0}(t) - (\mu_8 + \lambda_8)P_{S8}(t) + \lambda_8 P_{SF}(t) \tag{A.60}$$

P_{SF} is derived by the following lines:

$$P_{SF}(t + \Delta t) = a_{1F}P_{S1}(t) + a_{2F}P_{S2}(t) + a_{3F}P_{S3}(t) + a_{4F}P_{S4}(t)$$
$$+ a_{5F}P_{S5}(t) + a_{6F}P_{S6}(t) + a_{7F}P_{S7}(t) + a_{8F}P_{S8}(t) + a_{FF}P_{SF}(t) \tag{A.61}$$

By substituting expressions for availability matrix coefficients, we obtain:

$$P_{SF}(t + \Delta t) = (\mu_1\Delta t + \alpha_{15}\Delta t)P_{S1}(t) + (\mu_2\Delta t + \alpha_{26}\Delta t)P_{S2}(t)$$
$$+ \mu_3\Delta t P_{S3}(t) + \mu_4\Delta t P_{S4}(t) + (\mu_5\Delta t + \alpha_{54}\Delta t)P_{S5}(t) + \mu_6\Delta t P_{S6}(t)$$
$$+ (\mu_7\Delta t + \alpha_{74}\Delta t)P_{S7}(t) + [\mu_8\Delta t + (\alpha_{83} + \alpha_{84} + \alpha_{85} + \alpha_{86}$$
$$+ \alpha_{87})\Delta t]P_{S8}(t) + [1 - \Delta t(\mu_1 + a_{15} + \mu_2 + \alpha_{26} + \mu_3 + \mu_4 + \mu_5$$
$$+ \alpha_{54} + \mu_6 + \mu_7 + \alpha_{74} + \mu_8 + \alpha_{83} + \alpha_{84} + \alpha_{85} + \alpha_{86} + \alpha_{87})]P_{SF}(t) \tag{A.62}$$

Putting limit:

$$\lim_{\Delta t \to 0} \frac{P_{SF}(t + \Delta t) - P_{SF}(t)}{\Delta t} = \lim_{\Delta t \to 0} [(\mu_1 + \alpha_{15})P_{S1}(t) + (\mu_2 + \alpha_{26})P_{S2}(t) + \mu_3 P_{S3}(t)$$
$$+ \mu_4 P_{S4}(t) + (\mu_5 + \alpha_{54})P_{S5}(t) + \mu_6 P_{S6}(t) + (\mu_7 + \alpha_{74})P_{S7}(t)$$
$$+ (\mu_8 + \alpha_{83} + \alpha_{84} + \alpha_{85} + \alpha_{86} + \alpha_{87})P_{S8}(t)$$
$$- (\mu_1 + \alpha_{15} + \mu_2 + \alpha_{26} + \mu_3 + \mu_4 + \mu_5 + \alpha_{54}$$
$$+ \mu_6 + \mu_7 + \alpha_{74} + \mu_8 + \alpha_{83} + \alpha_{84} + \alpha_{85} + \alpha_{86} + \alpha_{87})P_{SF}(t)] \tag{A.63}$$

Finally, we get the differential equation for simulation in Simulink:

$$\frac{dP_{SF}(t)}{dt} = (\mu_1 + \alpha_{15})P_{S1}(t)$$
$$+ (\mu_2 + \alpha_{26})P_{S2}(t) + \mu_3 P_{S3}(t) + \mu_4 P_{S4}(t)$$
$$+ (\mu_5 + \alpha_{54})P_{S5}(t) + \mu_6 P_{S6}(t) + (\mu_7 + \alpha_{74})P_{S7}(t)$$
$$+ (\mu_8 + \alpha_{83} + \alpha_{84} + \alpha_{85} + \alpha_{86} + \alpha_{87})P_{S8}(t)$$
$$- (\mu_1 + \alpha_{15} + \mu_2 + \alpha_{26} + \mu_3 + \mu_4 + \mu_5 + \alpha_{54} + \mu_6 + \mu_7 + \alpha_{74}$$
$$+ \mu_8 + \alpha_{83} + \alpha_{84} + \alpha_{85} + \alpha_{86} + \alpha_{87})P_{SF}(t) \tag{A.64}$$

We tried to solve this system of differential equations in the same way as in the case of reliability, but we got the problem "out of memory". However, this should not discourage someone with a more powerful computer from trying to find the solution in the same way. In our computer, Matlab took up 25 % of processor power and almost the entire RAM.

A.3 Solving Differential Equations for Availability in Simulink

We attempted to solve the system in A.2 by simplifying the problem by introducing numbers for the numerical solution.

The first problem was the fact that Simulink can only be used for numerical values of constants, which is not a generalized solution. So, the solution can only be applied for specific case.

In order to solve the problem in the Simulink environment, we assumed that:

- every $\lambda_i = 1$,
- $\mu_i = 1$, and
- every summation and interaction weights are equal to 1.

In addition, we assumed unit step input, which is standard testing input in many applications.

Figure A.1 shows the Simulink model for the solution of the system. It is too large and some characters cannot be so big to be visible due to reduced size in book format.

Figures A.2 and A.3 show details of the model's make-up.

In Fig. A.2, we can see a great summation symbol, which has inputs from all availabilities with the corresponding λ-s and the feedback from the solution. This represents the right side of the differential equation and the integrator integrates in order to obtain the solution.

Similar method was used for other equations as well.

Figure A.3 illustrates a model for differential equations for P_{S4}, and is only a magnified part of Fig. A.1, which is also the case with Fig. A.1.

Figure A.3 is introduced in order to get clear picture about model. Bold circle shows output for P_{S4}, which is a scope—graphical representation of the output function.

As expected, the result (see Fig. A.4) for such a simplified case is an exponential function.

However, if we assume that:

- $\lambda_i = 0.01$,
- $\mu_i = 0.01$,
- $\alpha_{ij} = 0.005$,

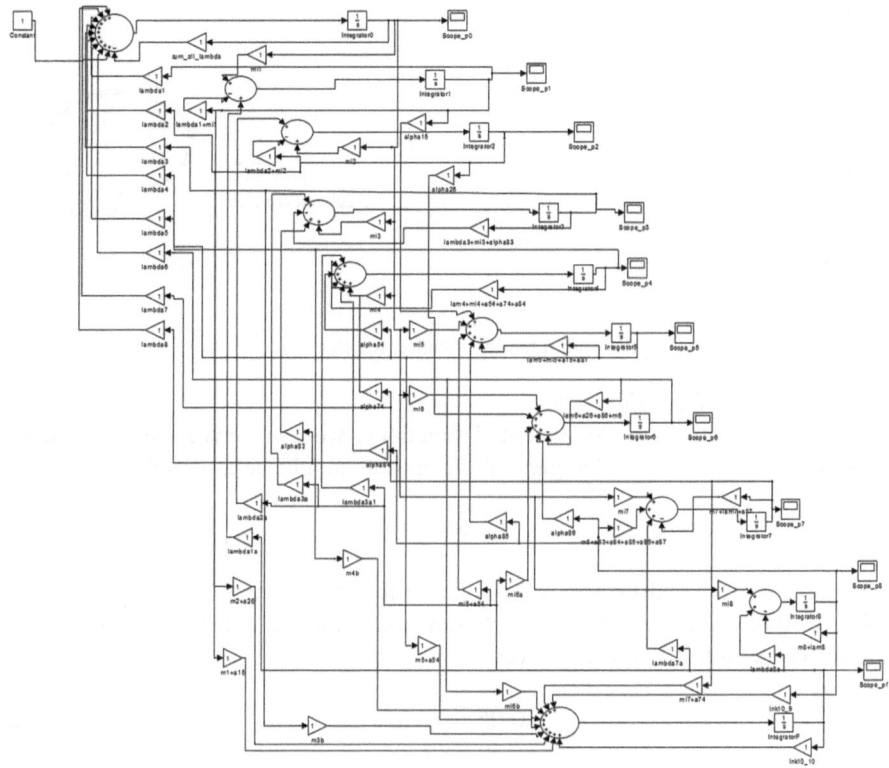

Fig. A.1 The Simulink solution for the entire system in availability case

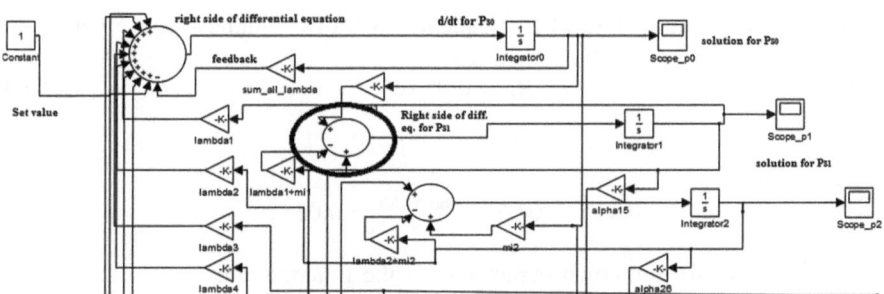

Fig. A.2 A part of the solution, which shows general structure of the solution to differential equations

**Sumation of right
side of the differential
equation**

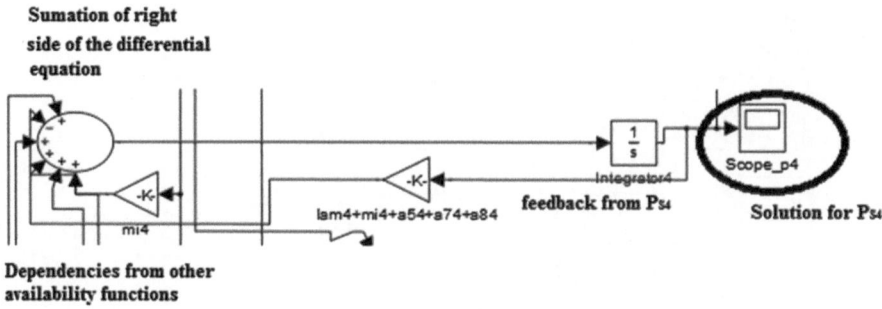

**Dependencies from other
availability functions**

Fig. A.3 A part of the solution—differential equation for P_{S4}

we get results from Fig. A.5.

Since the solutions are exponential functions, we can derive the approximate solution from the graphs. The general expression should be as follows:

$$P_{SF} = C \cdot e^{b \cdot t} \tag{A.65}$$

To obtain the actual values of the solution for P_{SF}, we can use an easily visible point (6, 0.05) and one randomly chosen point, i.e. (8, 0.115). By substituting bracket values into (A.65), we have:

$$0.05 = C \cdot e^{6b} \tag{A.66}$$

and

$$0.115 = C \cdot e^{8b} \tag{A.67}$$

When solving the system of two equations with two unknowns, we get:

$$C = 0.05 \cdot e^{-6b} \tag{A.68}$$

and

$$0.115 = 0.05 \cdot e^{-6b} \cdot e^{8b} = 0.05 \cdot e^{2b} \tag{A.69}$$

The solutions for unknown parameters of the function are:

$$b = 0.5 \cdot \ln(0.115/0.05) = 0.41645 \tag{A.70}$$

and

$$C = 0.00411 \tag{A.71}$$

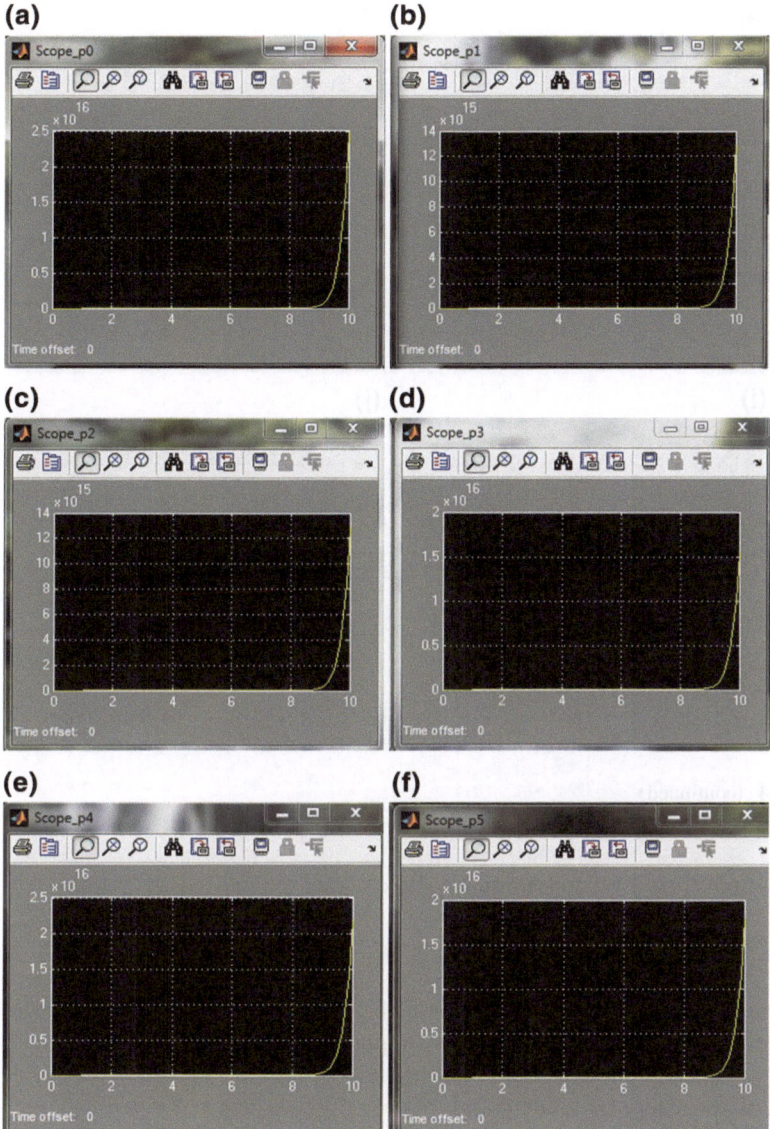

Fig. A.4 Graphs for solutions in simplified case: from P_{S0} (**a**) to P_{SF} (**j**)

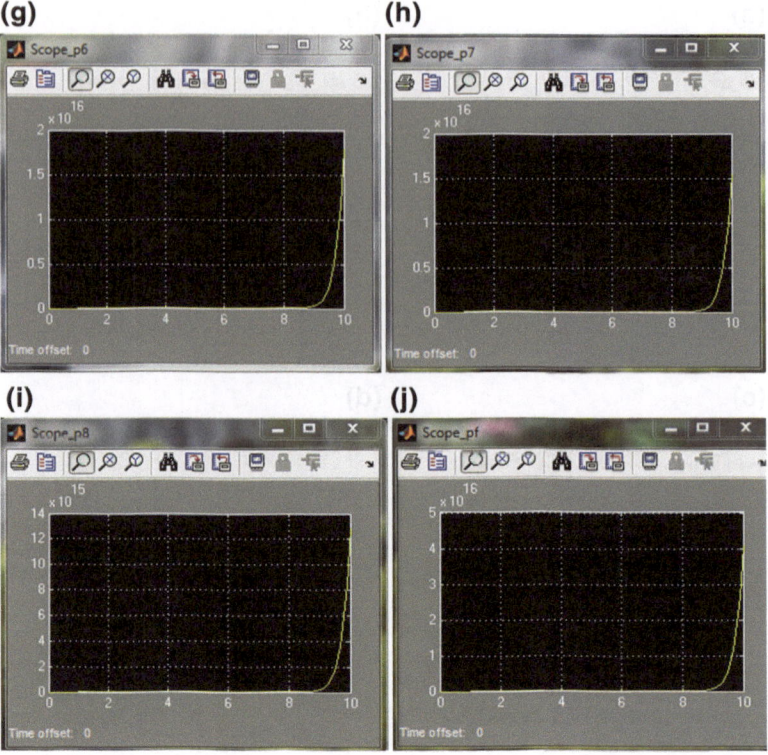

Fig. A.4 (continued)

The final form of the solution for exponential expression is:

$$P_{SF} = 4.11 \cdot e^{0.41645 \cdot t} \cdot 10^{-3} \tag{A.72}$$

Similar can be executed for all availability functions (probabilities). However, there are some drawbacks. The most important is connectivity with a real-world system. Namely, simulation time in Simulink is not the same as in real availability functions. In reality, seconds could be days, months or even years.

The second case is closer to reality, but the exact real-world values should still be obtained from the manufacturer.

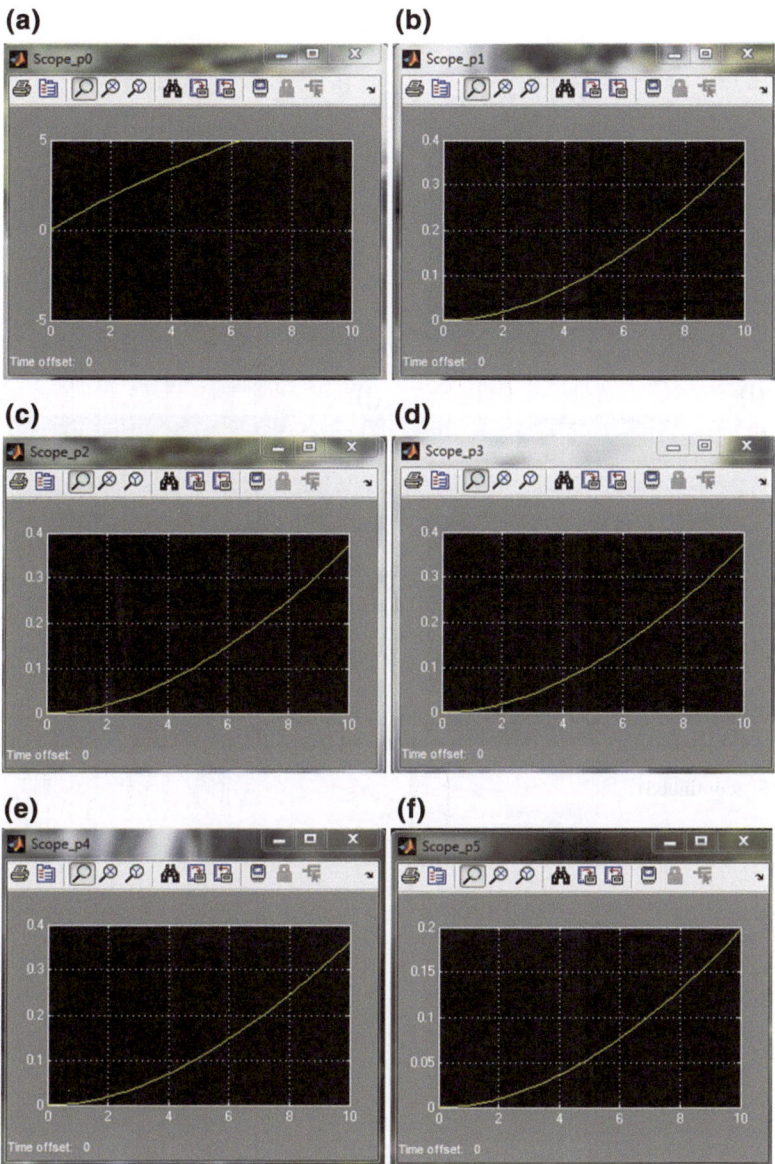

Fig. A.5 Graphs of the solutions for the more complex assumptions: P_{S0} (**a**), P_{S1} (**b**), P_{S2} (**c**), P_{S3} (**d**), P_{S4} (**e**), P_{S5} (**f**), P_{S6} (**g**), P_{S7} (**h**), P_{S8} (**i**), and P_{SF} (**j**)

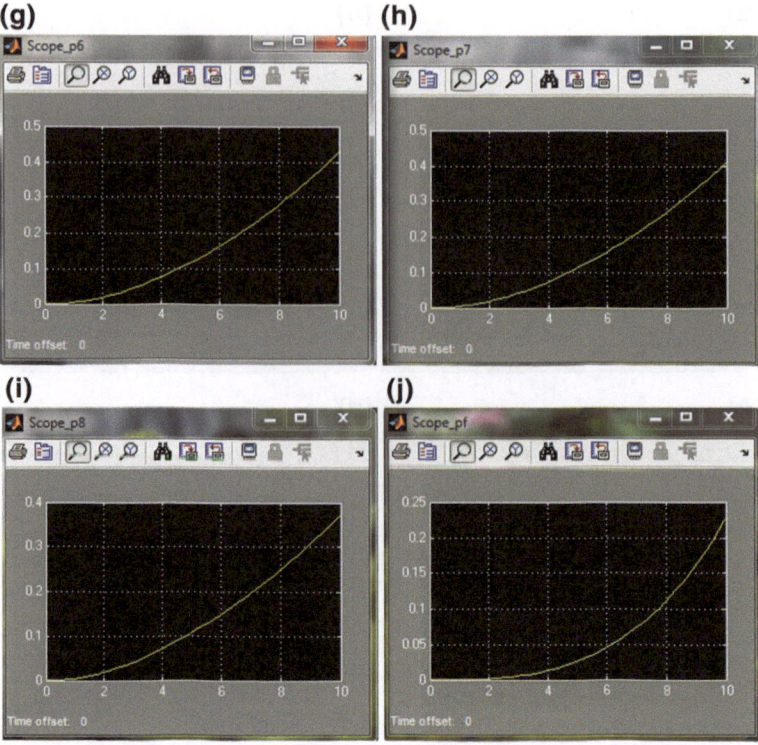

Fig. A.5 (continued)

References

1. Kuzmanić I, Vujović I, Šoda J (2014) Damage detection in materials based on computer vision wavelet algorithm. In: Öchsner A, Altenbach H (eds) Design and computation of modern engineering materials, Springer, Berlin, p 157–186
2. Dresbach C, Dressler U, Gussone J, Reh S (2013) Calculation of effective young's modulus distribution from electron backscatter diffraction results for stochastic analyses in aerospace applications. In: Abstract book of 8th, International conference on advanced computational engineering and experimenting. Madrid, Spain, pp 22, 1–4 July 2013
3. Lala PK (2001) Self-checking and fault-tolerant digital design. Academic Press, New York
4. Faulin J, Juan AA, Martorell S, Ramírez-Márquez JE (eds) (2010) Simulation methods for reliability and availability of complex systems. Springer, London
5. Rodriguez J, Remack K, Gertas J, Wang L, Zhou C, Boku K, Rodriguez-Latorre J, Udayakumar KR, Summerfelt S, Moise T, Kim D, Groat J, Eliason J, Depner M, Chu F (2010) Reliability of ferroelectric random access memory embedded within 130 nm CMOS. In: 2010 IEEE international reliability physics symposium (IRPS). Anaheim, USA, 2–6 May 2010
6. Kontoleon JM, Stergiou A (1991) Reliability analysis and design of a fault-tolerant random access memory system. Microelectron Reliab 31:1063–1067
7. Srinivasan J, Adve SV, Bose P, Rivers JA (2004) The case for lifetime reliability-aware microprocessors. in: proceedings of the 31st international symposium on computer architecture (ISCA-04). http://rsim.cs.illinois.edu/Pubs/srinivasan_isca04.pdf. Accessed 12 Aug 2013
8. Srinivasan J, Adve SV, Bose P, Rivers J, Hu CK, (2003) RAMP: a model for reliability aware microprocessor design. IBM research report, RC23048 (W0312-122) http://www.cs.utexas. edu/~hestness/papers/srinivasan-rampdetail.pdf. Accessed 13 Aug 2013
9. Shin J (2008) Lifetime reliability studies for microprocessor chip architecture. PhD Thesis, University of Southern California, Faculty of the Graduate School. http://digitallibrary.usc. edu/cdm/ref/collection/p15799coll127/id/108618. Accessed 12 Aug 2013
10. Li Q, Patel U (2005) Enabling memory reliability, availability, and serviceability features on dell poweredge servers. Dell Power Solutions 8(2005):1–4
11. Čoko M (2013) Analiza pouzdanosti i dostupnosti kompjuterskih sustava u Matlabu (Analysis of reliability and availability of the computer system in Matlab). Master's thesis, University of Split, Faculty of Maritime Studies
12. Louit D, Pascuall R, Banjevic D, Jardine AKS (2011) Optimization models for critical spare parts inventories—a reliability approach. J Oper Res Soc 62(2011):992–1004. http://web.ing. puc.cl/~rpascual/mispapers/louit10.pdf. Accessed 12 Aug 2013
13. Šoda J, Beroš SM, Kuzmanić I, Vujović I (2013) Discontinuity detection in the vibration signal of turning machines. In: Öchsner A, Altenbach H (eds) Experimental and numerical investigation of advanced materials and structures. Advanced structured materials. Springer, Heidelberg, pp 27–54
14. Mallat S (2009) A wavelet tour of signal processing. Academic Press, New York

I. Kuzmanić and I. Vujović, *Reliability and Availability of Quality Control Based on Wavelet Computer Vision*, SpringerBriefs in Electrical and Computer Engineering, DOI 10.1007/978-3-319-13317-1

15. Daubechies I (1992) Ten Lectures on Wavelets. Society for industrial and applied mathematics. Philadelphia, USA
16. Christopher H, Walnut DF (2006) Fundamental papers in wavelet theory. Princeton University Press, London
17. Mitsuru O (1984) Software reliability analysis models. IBM J Resear Develop 28(4):428–443
18. Sahner RA, Trivedi K, Puliafito A (2012) Performance and reliability analysis of computer system: an example-based approach using the sharpe software package. Springer, Heidelberg
19. Hassett TF, Dietrich DL, Szidarovszky F (1995) Time-varying failure rates in the availability and reliability analysis of repairable systems. IEEE Tran Reliab 44(1):155–160
20. Suesut T, Numsomran A, Tipsuwanporn V (2004) Vision-based network system for industrial applications. Int J Comp Syst Sci Eng 3(1):22–26
21. Herakovic N (2010) Robot vision in industrial assembly and quality control processes. In: Ude A (ed) Robot vision. InTech, Rijeka, pp 501–534
22. Crowley JL, Hall D, Emonet R (2007) Autonomic computer vision systems. In: Proceedings of the 5th international conference on computer vision systems (ICVS 2007). http://biecoll.ub.uni-bielefeld.de. Accessed 1st Sept 2013
23. Wang Z, Bovik AC, Sheikh HR, Simoncell EP (2004) Image quality assessment: from error visibility to structural similarity. IEEE Tran Imag Process 13(4):600–612
24. Algorithm for image quality assessment (2004) http://www.cns.nyu.edu/~lcv/ssim/. Accessed 4th Jun 2013
25. JND metrics software (2014) http://videoclarity.com/PDF/ClearViewDataSheet.pdf. Accessed 2nd Jul 2013
26. Akbulut A, Ilk HG, Ari F (2005) Design, availability and reliability analysis on an experimental outdoor FSO/RF communication system. In: Proceedings of 2005 7th International Conference. Transparent optical networks. Barcelona, Spain, vol. 1, pp 403–406, 3–7 July 2005
27. Tal R (2009) Real-time approaches to computer vision. report, elder laboratory—human & computer vision. http://elderlab.yorku.ca/~rontal/cse5441/project_rt_vis.pdf. Accessed 4th Mar 2013
28. Raji A, Alamutu A (2005) Prospects of computer vision automated sorting systems in agricultural process operations in Nigeria. Agric Eng Int: CIGR J Sci Resear Develop 7 (2):1–12
29. Wu D, Sun DW (2012) Colour measurements by computer vision for food quality control—a Review. Trends Food Technol 29(1):5–20
30. Narendra VG, Hareesh KS (2010) Prospects of computer vision automated grading and sorting systems in agricultural and food products for quality evaluation. Int J Comput Appl 1(4):1–9
31. Carlson J, Murphy RR (2003) Reliability analysis of mobile robots. In: Proceedings of ICRA'03 IEEE international conference on robotics and automation, 14–19 Sept 2003. Taipet, Taiwan, vol. 1, pp 274–281
32. Tavner PJ, Xiang J, Spinato F (2007) Reliability analysis for wind turbines. Wind Energ 10:1–18
33. Young RB (2003) Reliability transform method. Master's thesis, Virginia Polytechnic Institute and State University, Blacksburg
34. Vujović I, Šoda J, Kuzmanić I (2012) Cutting-edge mathematical tools in processing and analysis of signals in marine and navy. Trans Marit Sci 1(1):35–48
35. Gabor D (1946) Theory of communication. J Inst Electr Eng 93(26):429–457
36. Herley C, Kovačević J, Ramchandran K, Vetterli M (1993) Tilings of the time-frequency plane: construction of arbitrary orthogonal bases and fast tiling algorithms. IEEE Tran Sig Process 41(12):3341–3359
37. Antoniou A (2006) Digital signal processing—signals. Systems and filters. McGrow-Hill, New York
38. Proakis JG, Manolakis DK (2007) Digital signal processing. Prentice-Hall, New York
39. Jansen M, Oonincx P (2005) Second generation wavelets and applications. Springer, London
40. Strang G, Nquyen T (1997) Wavelets and filter banks. Wellesly—Cambridge Press, Boston

41. Kingsbury NG, Magarey JFA (1997) Wavelet transforms in image processing. In: Proceedings of first European conference on signal analysis and prediction, Prague, Czech Republic, 24–27 June. Birkhäuser, pp 23–24. http://www-sigproc.eng.cam.ac.uk/∼ngk/publications/ngk97b.zip. Accessed 18 Aug 2011

42. Lina JM, Gagnon L (1995) Image enhancements with symmetric Daubechies' wavelets. SPIE Aerosense: Wavelet Appl Signal Image Process III 2569(1):196–207

43. Vujović I, Šoda J, Beroš SM (2012) Time-frequency methods in maritime surveillance systems. Precious Sea 59(5-6):254–265

44. Anaraki MS, Dong F, Hirota K, Nobuhara H (2007) A multidirectional multiresolution transform for image representation. In: 8th international symposium on advanced intelligent systems ISIS 2007, Sokcho-City, Korea, 5–8 Sept. http://isis2007fuzzy.or.kr/submission/upload/A1331.pdf. Accessed 22 Aug 2011

45. Candés E, Demanet L, Donoho D, Ying L (2006) Fast discrete curvelet transforms. Multiscale Model Simul 5(3):861–899

46. Donoho D (1999) Wedgelets: nearly-minimax estimation of edges. Ann Stat 27(3):859–897. doi:10.1214/aos/1018031261

47. Melchior P, Meneghetti M, Bartelmann M (2007) Reliable shapelet image analysis. Astron Astrophys 463(3):1215–1225. doi:10.1051/0004-6361:20066259

48. Pennec E, Mallat S (2005) Sparse geometric image representations with bandelets. IEEE Tran Image Process 14(4):423–438. doi:10.1109/TIP.2005.843753

49. Wu B, Nevatia R (2005) Detection of multiple, partially occluded humans in a single image by Bayesian combination of edgelet part detectors. In: 10th IEEE international conference on computer vision, Beijing, China, 17–20 Oct, Springer, pp 90–97. doi: 10.1109/ICCV.2005.74

50. Selesnick IW, Baraniuk RG, Kingsbury NG (2005) The dual-tree complex wavelet transform. IEEE Sign Process Mag 22(6):123–151. doi:10.1109/MSP.2005.1550194

51. Kingsbury N (1998) The dual-tree complex wavelet transform: a new efficient tool for image restoration and enhancement. In: Proceedings of European signal processing conference, EUSIPCO 98, Rhodes, 8-11 Sept, University of Athens, pp 319–322

Index

A
Action phase, 11
Availability, 5

C
Compression, 8, 9

D
2D-DWT, 8
Denoising, 8
Diagnostic tools, 1

E
Edge detector, 8, 17
Electronic devices, 20

F
Failure intensity, 20, 21

H
Hardware/environment parameter, 15

I
Illumination, 2
Image quality, 2
Influencing parameters, 15
Intensity of failure, 5, 21, 22, 32
Intensity of repairs, 5, 7, 31, 32

M
Markov chains, 2
Markov model, 2, 18
Matlab, 27, 28, 39, 41, 43, 45, 46, 48, 49, 51, 56
MTBF, 2
MTTR, 5

N
Network reliability, 2
Noise, 11
Noisy conditions, 11
Noisy data, 2

O
Observe phase, 11

Q
Quality control systems, 1

R
Reliability analysis, 1
Reparable system, 31
Robot vision, 2

S
Simulink, 39, 41, 55–57, 60
State transition, 19
Stochastic, 2

© The Author(s) 2015
I. Kuzmanić and I. Vujović, *Reliability and Availability of Quality Control Based on Wavelet Computer Vision*, SpringerBriefs in Electrical and Computer Engineering, DOI 10.1007/978-3-319-13317-1